12th Anglo-French Physical Acoustics Conference

(AFPAC 2013)

Journal of Physics: Conference Series Volume 498

Frejus, France
16 – 18 January 2013

ISBN: 978-1-63266-723-6
ISSN: 1742-6588

Printed from e-media with permission by:

Curran Associates, Inc.
57 Morehouse Lane
Red Hook, NY 12571

Some format issues inherent in the e-media version may also appear in this print version.

Printed by Curran Associates, Inc. (2014)

Institute of Physics
Dirac House, Temple Back
Bristol BS1 6BE UK

Phone: 44 1 17 929 7481
Fax: 44 1 17 920 0979

techtracking@iop.org

Additional copies of this publication are available from:

Curran Associates, Inc.
57 Morehouse Lane
Red Hook, NY 12571 USA
Phone: 845-758-0400
Fax: 845-758-2634
Email: curran@proceedings.com
Web: www.proceedings.com

TABLE OF CONTENTS

Preface

The Anglo-French Physical Acoustics Conference (AFPAC) had its 12th annual meeting in Villa Clythia, Fréjus, France, from 16th to 18th January 2013. This series of meetings is a collaboration between the Physical Acoustics Group (PAG) of the Institute of Physics and the Groupe d'Acoustique Physique, Sous-marine et UltraSonore (GAPSUS) of the Société Française d'Acoustique. This year, attendees got the opportunity to see the French Riviera with its Mediterranean vegetation covered by a nice thick snow layer.

The participants heard 34 excellent oral presentations and saw 3 posters covering an exciting and diverse range of subjects and of frequencies, from ultrasonic wave propagation in chocolate to metamaterials applied to seismic waves for protecting buildings. Among them, invited talks were given by Pr F. A. Duck (*Enhanced healing by ultrasound: clinical effects and mechanisms*), Pr. J.-C. Valiére, who actually gave two invited talks (*1. Measurement of audible acoustic particle velocity using laser: Principles, signal processing and applications, 2. Acoustic pots in ancient and medieval buildings: Literary analysis of ancient texts and comparison with recent observations in French churches*), Dr. P. Huthwaite (*Ultrasonic imaging through the resolution of inverse problems*), Dr. X. Lurton (*Underwater acoustic systems on oceanographic research vessels: principles and applications*), Dr. S. Guenneau (*From platonics to seismic metamaterials*).

For the fifth consecutive year AFPAC is followed by the publication of its proceedings with 12 peer-reviewed papers which cover the most recent research developments in the field of Physical Acoustics in the UK and France.

Alain Lhémery (CEA, France) and Nader Saffari (UCL, United Kingdom)

Journal of Physics: Conference Series **498** (2014) 011001 doi:10.1088/1742-6596/498/1/011001

12th AFPAC – Villa Clythia, Fréjus (French Riviera), the 17th of January 2013

Peer review statement

All papers published in this volume of *Journal of Physics: Conference Series* have been peer reviewed through processes administered by the proceedings Editors. Reviews were conducted by expert referees to the professional and scientific standards expected of a proceedings journal published by IOP Publishing.

Journal of Physics: Conference Series **498** (2014) 012001 doi:10.1088/1742-6596/498/1/012001

Can airborne ultrasound monitor bubble size in chocolate?

N Watson[1], T Hazlehurst[1], M Povey[1], J Vieira[2], R Sundara[2], J-P Sandoz[3]

[1]School of Food Science and Nutrition, University of Leeds, UK.
[2]Nestlé Product Technology Centre, York, UK.
[3]Institute of Applied Microtechnology, HE-ARC, Switzerland.

E-mail: N.J.Watson@Leeds.ac.uk

Abstract. Aerated chocolate products consist of solid chocolate with the inclusion of bubbles and are a popular consumer product in many countries. The volume fraction and size distribution of the bubbles has an effect on their sensory properties and manufacturing cost. For these reasons it is important to have an online real time process monitoring system capable of measuring their bubble size distribution. As these products are eaten by consumers it is desirable that the monitoring system is non contact to avoid food contaminations. In this work we assess the feasibility of using an airborne ultrasound system to monitor the bubble size distribution in aerated chocolate bars. The experimental results from the airborne acoustic experiments were compared with theoretical results for known bubble size distributions using COMSOL Multiphysics. This combined experimental and theoretical approach is used to develop a greater understanding of how ultrasound propagates through aerated chocolate and to assess the feasibility of using airborne ultrasound to monitor bubble size distribution in these systems. The results indicated that a smaller bubble size distribution would result in an increase in attenuation through the product.

1. Introduction

There are many different varieties of chocolate bars available to consumers with one of the most popular being aerated chocolate bars. Aerated chocolate bars consist of a solid bar with the inclusion of many bubbles. The gas in the bubbles is usually a carbon dioxide mix and is added to the bars whilst the chocolate is in a liquid state. Once the chocolate solidifies, the nitrous oxide disperses in the atmosphere to be replaced by atmospheric gases. An important property of aerated chocolate bars is the volume fraction of gas and the bubble size distribution. These features have a significant effect on the production cost, sensory attributes and consumer perception of the product, therefore the bubble size distribution is monitored during production. Current techniques to monitor bubble size in these products consist of removing samples from the production line, breaking the bar to expose the bubbles and making a visual observation of the bubble size distribution. This method is far from ideal as the classification of bubble size is subjective, and can often only identify undesirable bubble sizes in products long after the fault has occurred. There is an industry need for online real time techniques capable of monitoring the bubble size in these products to optimise the manufacturing process, ensure product consistency and minimise production waste. Acoustic and optical techniques do exist to

[1] To whom any correspondence should be addressed.

measure bubbles sizes in systems such as the fermentation process [1] but the majority of these have only been utilised in liquid systems. Acoustic techniques are well developed and employed for measuring dispersed phase size of droplets in emulsions [2, 3] although it should be remembered that systems with bubbles are very different. Existing acoustic techniques comprise of either listening to the sound of moving or collapsing bubbles (in a liquid system) or transmitting ultrasound through the systems [4-7]. Most active transmission technique require multiple measurements on the system under investigation and can require complex data analysis rendering them unsuitable for an online measurement system on a fast moving production line due to the time required to make multiple measurements and process the data.

For process monitoring in the food industry there is a need for non contact and non invasive techniques as contact could risk contamination of the food and potentially impact the process. This would make most acoustic techniques unsuitable as the transducers are generally in contact with the object under inspection. One solution would be to use a non contact air coupled system. This technique works by transmitting an acoustic pressure wave from one transducer through air, the product under investigation and through air again before the wave is detected by the receiver. The properties of the tested food will affect the received signal so it is possible to infer information about this product from the characteristics of the received pulse. Airborne ultrasound systems have being used to detect foreign bodies in food products [8] but to the authors knowledge no studies have being performed using the technique to monitor bubble sizes in aerated chocolate products.

Computational modelling techniques exist to study how ultrasound will propagate through systems such as aerated chocolate bars and can be used to study the effects of varying experimental parameters such as transducer frequency and location. If a well validated model is constructed it can be used to optimise the design process of new instrument development by performing a large number of virtual experiments which would be costly both in time and monetary terms to perform with actual experiments.

The purpose of this work is to use a combined experimental and modelling approach to assess the feasibility of using airborne ultrasound to monitor bubble size distributions in aerated chocolate bars. This was achieved by constructing two model two dimensional aerated chocolate products from straws and comparing the acoustic results with those from a COMSOL model with the same geometry. Straws were chosen as they have the same cross sectional area across their length so a direct comparison can be made to the 2D computational model. Experimental measurements were then made on two aerated chocolate bars known to have different bubble size distributions. These results were compared with two dimensional simulation results in which the geometry is a cross sectional area with a bubble size distribution representative of the corresponding aerated bar. The representative cross sectional area was determined by slicing each of the aerated chocolate bars, photographing the exposed area and using image analysis to calculate the bubble size distribution. Since the surface topography and material property is known to affect airborne ultrasound propagation this was also investigated experimentally in order to understand ultrasound propagation in these systems and to determine if the variations in ultrasound results are dominated by the bubble size distribution or other properties of the aerated product. This information will help determine the feasibility of developing an airborne ultrasound system for online real time monitoring of bubble size distribution.

Ultimately this research will provide valuable information into the viability of using airborne ultrasound as a method to monitor bubble size distribution in aerated chocolate bars online in real time.

This paper will start with a description of the experimental system and how it was used followed by how the computational model was developed. Where possible the model was constructed to represent the experiment as closely as possible. The results and discussion will be combined into a single section. First the experiment and simulation results will be presented and compared for the straws. Next the experimental and simulation results will be presented and compared for the actual aerated chocolate. The final section of the results and discussion will experimentally study the affect of the topography and surface condition on aerated chocolate bars.

2. Experimental methods

2.1 Airborne ultrasound system

For all measurements a bespoke airborne ultrasound system was used. This system can be seen in Figure 1. The set up comprises of a pair of Ultran non contact transducers (1). Three different pairs of transducers were used. One with a central frequency of 50 KHz (NCG50-S25), one with a central frequency of 140 KHz (NCG140-S38) and one with a central frequency of 200 KHz (NCG200-S38). Each pair of transducers was separated by a distance of 76 mm as this is the focal distance of all transducers. The transducers were attached to a robotic stage so they could be automatically moved over the aerated chocolate or straws. The straws or aerated chocolate were place at the centre distance between the transducers using a mould with the bottom cut away (2). This was to ensure that the sound wave only travelled though air and the sample under inspection. Both transducers were attached to a handy scope HS3 USB oscilloscope (3). This was controlled by purpose built Labview software. The software was used to send and receive the ultrasound pulse and analyse the received signal. The software also controlled the robotic stage. As ultrasound is highly attenuated in air both the transmitted and received pulse were amplified using an A 303 high voltage amplifier and modulator (A A lab systems, Israel) (4) for the transmitted pulse and THS4022 board (Texas instruments, USA) (5) for the received signal.

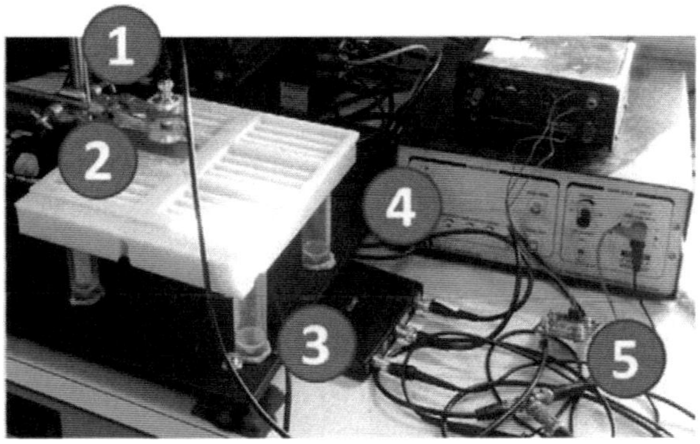

Figure 1. Airborne ultrasound system.

2.2 Experiments performed

Three sets of experiments were performed using the airborne ultrasound system:

1. Small and large straw configuration

2. Aerated chocolate bars with small and large bubble size distribution

3. B Scan along the length of an aerated bar with different surface conditions

Each set of experiments measurements were made with all three pairs of airborne ultrasound transducers. For each transducer the transmitting frequency was selected as the central frequency of the transducer pair (50, 140, 200 KHz). The transmitted pulse was 50 μs seconds long and had a sinusoidal shape function with a period twice that of the transmitted pulse duration. For all experiments only the attenuation of the ultrasound pulse was measured. Initially a measurement was made through air alone so that all the attenuation results could be presented relative to air. For this calibration measurement the transmitted amplitude was set to 0.01 V. When the straws or aerated chocolate was placed between the transducers the transmitting amplitude was increased to 6V. Once the measurement was made through air the received pulse was saved in the software. The amplitude of all other received pulses from the experiments was increased so that they matched this reference amplitude and the attenuation could be calculated. For the final set of experiments the pair of airborne transducers was scanned along the length of an aerated chocolate bar using the bespoke robotic platform. The purpose of these experiments was to determine the effect that the shape of the chocolate would have on the ultrasound attenuation. This is important as few chocolate bars are flat so the effect the shape has on ultrasound attenuation must be studied. If the shape has more of an effect than the bubble size distribution it may make the development of a process monitoring system difficult. In addition to studying the effect of aerated chocolate bar topography the surface condition of the chocolate may affect the attenuation. This was studied by placing small chocolate crumbs or an oil layer on the top and bottom of the bar for a particular area and then scanning the airborne transducers over the length of the bar. For these experiments the transducers were positioned in the center width of the bar and scanned along the length by 80 mm. Figure 8 A and B show the region of the bar scanned. For each scan a measurement was made every 0.5 mm.

2.3 Result representation

There are many different methods used to analyse acoustic propagation through a system such as speed of sound or signal frequency content but it was decided to study only the attenuation through the straws or aerated system. This method was chosen as the ultimate aim is to develop an online real time system so we wanted to minimize the amount of signal analysis required. This was also the rationale as to why we only used the single central frequency for each transducer. Spectroscopy techniques are capable and successful at measuring bubble size distributions but they require algorithms to compute the band width and the dispersive effects of each transducer and would not be suitable using the current system and may not be suitable for a real-time system with temporal changes. For all results presented in this paper the peak amplitude of the received pulse was measured. This was then compared to the amplitude of the received pulse which had only travelled through air so that the attenuation (relative to air) could be measured for each of the experiments using the equation below.

$$Attenuation\ (dB) = 20log10(600 \times peak\ amplitude\ difference)$$

The 600 represents the voltage increase for the transmitted pulse (0.01 to 6) and the peak amplitude difference is the difference in amplitude between the received pulse which has travelled through air and the pulse which had travelled through the straws or aerated chocolate bar. When measuring the peak amplitude of the received pulse it was important that this was measured on the section of the received wave which had travelled through the straws or aerated chocolate bar. This was achieved by windowing the received signal. Ten measurements were made for each experiment and the average was recorded. For each measurement the aerated chocolate bar was removed and replaced so the error associated with repeated changes in location could be observed. For the experimental results the attenuation on the figures is the average of the ten measurements and the error bars represent the standard deviation.

2.4 Image analysis of aerated chocolate

The aerated chocolate was sliced so that the bubble size distribution was exposed. This was done for two aerated chocolate bars, one which was known to include abnormally large bubbles and one which was known to include abnormally small bubbles. The exposed cross section area was photographed using a Canon 500D digital Camera and subsequent image analysis was performed using ImageJ [9]. The image was cropped so that only the area containing the bubbles remained; this was then enhanced and converted to a binary image. A water shedding function was employed so that the individual bubbles could be seen and the particle size function was used to calculate the average area of the bubbles and generate an outline of the bubbles which could be used in the COMSOL model. From the ImageJ results if we assume that the bubbles are spherical in size the average radius of the small bubble size distribution is 0.48 mm and the average radius of the large bubble size distribution is 0.56 mm.

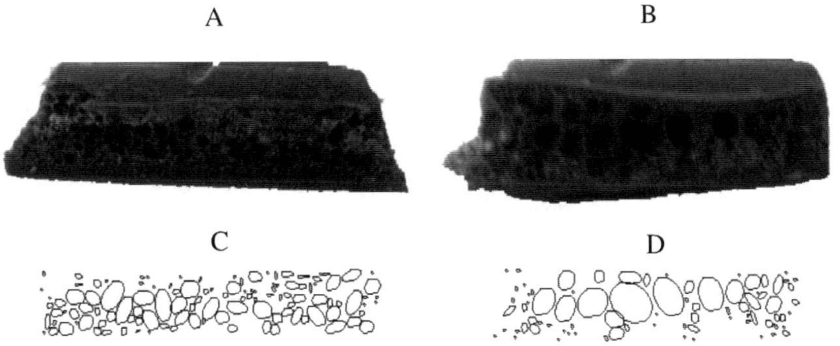

Figure 2. Aerated chocolate bar cross sections and representative bubble size distributions.

3. Computation modelling

3.1 Model definition

A computational model was created using COMSOL 4.3a to simulate how ultrasound propagates through aerated products. Four models were made, one for the small straws, one for the large straws, one for the aerated chocolate with a small bubble size distribution and one for the aerated chocolate with a large bubble size distribution. All models used the COMSOL transient pressure acoustics model and were created in two dimensions. Although the actual aerated chocolate bars would be more representative in a three dimensional model, the small element size and simulation time step required to accurately model the acoustic propagation in 3D would increase the simulation duration to an unacceptable time with the current computational setup.

3.2 Geometry

The Geometry of the COMSOL model can be seen in Figure 3 A, The horizontal line at the bottom represents the transmitting transducer and the horizontal line at the top represents the receiving transducer. Both of these lines have representative dimensions of the airborne ultrasound transducers used for the experiments. The arrow in this diagram represents the acoustic propagation path. The geometry of the straws can be seen in Figure 3 B and C. The size number and wall thickness of the straws was modelled with the dimensions used in the airborne experiments. The geometry for the aerated chocolate consisted of a rectangle with height 13 mm and width 44 mm (approximate dimensions of aerated chocolate bars used). The two dimensional images of the small and large bubble size distributions calculated from image analysis (Figure 2 C and 2 D) were placed in the centre of these rectangles. The two horizontal lines (representing transmitting and receiving transducers were separated by a distance of 76mm as was used in the experiment.

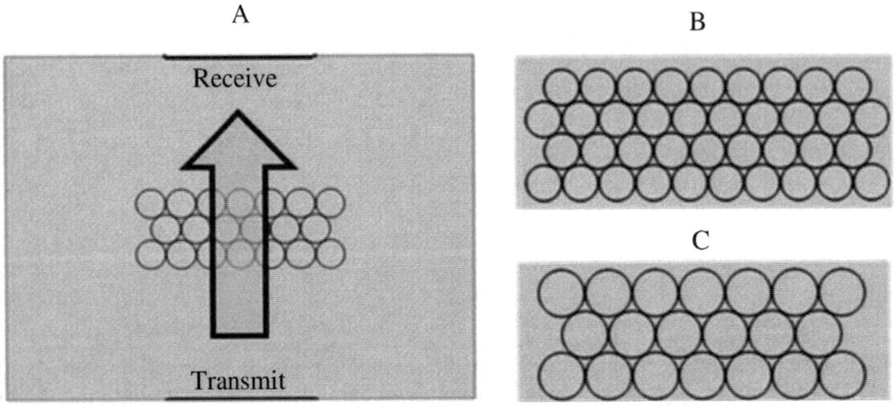

Figure 3. Computational model geometry, A: Model domain with transducers straws and air. B: Geometry of small straws, C: Geometry of large straws.

3.3 Acoustic wave propagation

To simulate the ultrasound propagation the bottom horizontal line experienced a plane wave radiation. Three different frequencies of this pulse were simulated; these were 50, 140 and 200 KHz. These were the same frequencies used in the airborne experiments. Each transmitted pulse was given a duration and shape identical to the experiments. The ultrasound pressure wave propagated through the air and aerated chocolate in the model and was detected at the top horizontal line in Figure 3 A, this represented the receiver transducers. The pressure was integrated across this line as is the case in acoustic transducers.

3.4 Material properties

The three materials used in the simulations were air, chocolate and paper (for the straws). The important information for the ultrasound propagation was the longitudinal speed of sound and the material's density. Air is a well defined property within the software so already had the required values. For chocolate and paper these material properties can be seen in Table 1. For the speed of sound of the chocolate a non aerated bar of the same chocolate was produced and the speed of sound was measured experimentally using a contact ultrasound system. The density of the chocolate was calculated by weighing an amount of known volume. The speed of sound and density of the paper straws was measured using similar techniques.

Table 1. Simulation material properties

	Longitudinal Speed of sound (m/s)	Density (Kg/m^3)
Chocolate	1500	1325
Paper	1325	800

3.5 Finite element mesh

The geometry of the system was discretised using a finite element mesh. A free triangular mesh was used for all models. An important aspect is the size and number of mesh elements utilised as this governs the accuracy, precision and time duration of the simulation. A trade off always exists to this extent. Adopting the method of Mylavarapu and Boddapati [10] the maximum element size used was the ultrasound pulse wavelength divided by ten. This results in ten elements for each cycle of the waveform. The wavelength was calculated from the speed of sound and the transmitted frequency used.

3.6 Solver configurations
All simulations used a single time dependent solver. The selection of the simulation time step is important as it controls the accuracy resolution and simulation duration. Mylavarapu and Boddapati [10] stated that the simulation time step should be less than ((wavelength/10)/speed of sound). This approach was used in this work. The longitudinal speed of sound used in this equation was chosen as the highest value for a given simulation (paper or chocolate) to give the smallest critical time step and ensure mathematical stability.

3.7 Data analysis
To ensure that the simulation results could be compared directly to the experiments the measurements were made in the same way. First a simulation was performed without the presence of the straws or aerated chocolate geometry and the maximum amplitude of the received pulse was recorded. For all subsequent simulations through the straws and the aerated chocolate the peak amplitude of the received pulse was recorded so that the attenuation is decibel relative to air could be calculated.

4. Results and discussion

4.1 Model 2D straw systems

Figure 4 display the results for attenuation through the small and large straw configurations at the three different frequencies. Generally the attenuation is greater at a higher frequency which is usual for acoustic results as attenuation increases with the propagating frequency squared. For all frequencies the attenuation was highest in the smaller straws. As the total height of the straws was the same the smaller straws consisted of more straws in the height of material. This resulted in an increased number of interfaces reflecting the sound and an overall increase in attenuation. The difference in attenuation between the small and large straws was not identical at all frequencies. At 50 KHz there was a difference of around 3 dB whereas at 140 and 200 KHz the differences were approximately 0.5 and 1 dB respectively. The results from the simulations with the straw geometries showed the same trends as from the experiment. The attenuation was found to increase with frequency and the small straws had the highest attenuation. Consistent with the experiments the difference in attenuation between the small and large straws was greatest for the 50 and 200 KHz transducers. It should be noted that the three different pairs of transducers used all had slightly different geometries which might influence some of the results. Quantitatively the attenuation was found to be higher in the experimental system and the attenuation difference between the small and large straw geometries were greater from the simulations.

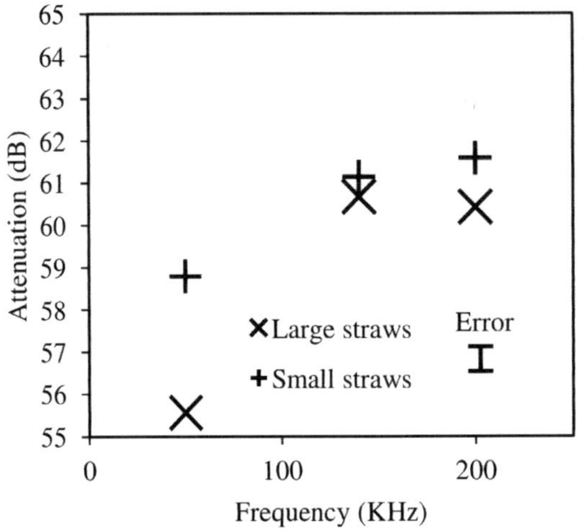

Figure 4. Airborne attenuation measurements on model systems of small and large straws.

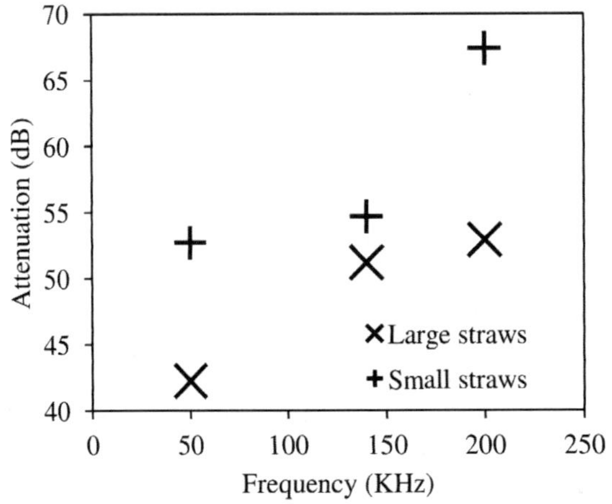

Figure 5. COMSOL attenuation simulations on model systems of small and large straws.

These results are very encouraging as they indicate that airborne ultrasound may distinguish between a small and large bubble size distribution in a model two dimensional system. Although inversion methods have not been used to calculate an actual bubble size distribution from the acoustic data the results do show that an indication of size can be taken from one simple ultrasound measurement. This would allow for quick online determination if a sudden change had occurred in the bubble size distribution. A system would require calibration with a known good bubble size distribution but calibration is common practice for online monitoring systems and the advantage of a processing monitoring system which can instantly identify a change in a product property such as bubble size has large commercial benefits. To understand how beneficial an airborne ultrasound system would be for monitoring bubble size distribution it is important to understand and investigate all factors of the product which will affect ultrasound propagation so that it can be determined how accurately can the system detect changes in the bubble size distribution. It should be noted that these experiments and simulations were only performed at two representative size distributions; more sizes and distributions should be studied to fully determine the capability of such a system.

The computational model predicts the main features of the experiments. The use of computational models is becoming increasingly popular since they offer many benefits. Once a model is developed and validated with suitable experiments such as those describe here, it can then be used to aid the understanding of the science of the system under question as well as aid the development process of a process monitoring system. The model developed during this research could be used to simulate a wide range of frequencies and transducer geometries to find the most suitable one for studying the aerated product under inspection. The model can also be used to optimise parameters in the system such as pulse repetition frequency and precise transducer location. These simulation results could then be used to minimise the time and cost required to develop the process monitoring system and increase the possibility of producing a successful commercial system.

4.2 Chocolate with different bubble size distributions

Experimental results from the airborne experiments on aerated chocolate bars with a small and large bubble size distribution show that attenuation increases with frequency and is higher in the bar with the small bubble size distribution. These results are consistent with the measurements for the model straw system. The difference in attenuation between the small and large bubble size distributions was found to be larger for the lower frequencies, with a generally greater differential than for the straws experiments. Image analysis revealed that the small and large bubble sized aerated bars had approximately the same volume fraction of gas. For the small size distribution this would result in more scattering objects (bubbles) in the active acoustic area resulting in the observed increase in attenuation.

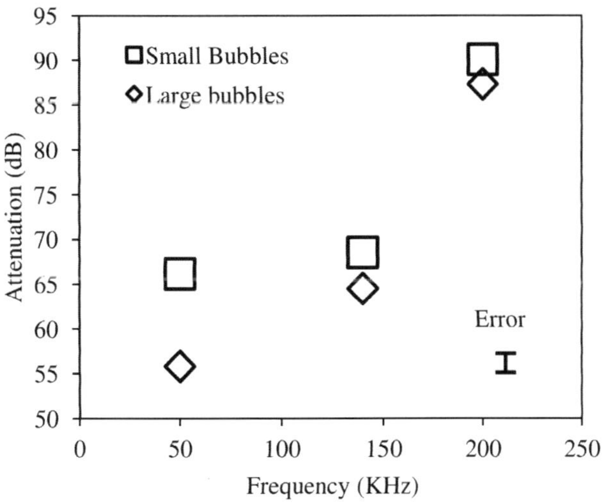

Figure 6. Experimental attenuation results from airborne ultrasound system using aerated chocolate bars with a small and large bubble size distribution.

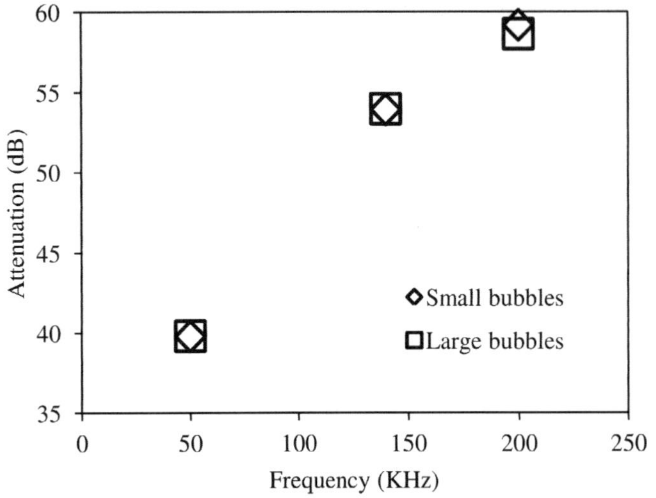

Figure 7. Simulation attenuation results from airborne ultrasound system using aerated chocolate bars with a small and large bubble size distribution.

Simulation results for models with representative small and large bubble size distributions (taken from image analysis of bars) indicate that attenuation increases with frequency although very little difference was observed for the two different bubble sizes. For all frequencies the attenuation was greater for the small bubble size distribution but this was only really noticeable for the simulations at 200 KHz. This difference was small compared with all other experimental and simulation results presented in this research.

The COMSOL model used was a two dimensional representation of the system and this could account for the difference between the experimental and simulations results. The straws experiment and models had almost identical geometry and comparable results. The geometry used in the models of the aerated chocolate bars were much simpler than the actual real life system as the image analysis only generated a representation of the bubble size distribution and contained far fewer bubbles than are known to exist in the actual products. The bubble area fraction from the image analysis was 35.5% for the small bubbles and 34% for the large bubbles. For the actual aerated products it is known that the aerated area size fraction is around 50%. This difference is accounted for by the lack of small bubbles detected by the image analysis technique and it may be that these small bubbles have the greatest effect on attenuation. With more accurate image recording and processing it will be possible to build more accurate models capable of predicting the experimental system.

4.3 Effect of chocolate surface condition

B scans along the length of the aerated chocolate bars were performed to study the effect of the topography of the products. The surface condition was also changed on certain sections of the bars by adding either chocolate crumbs or an oil layer. The initial results showed that the attenuation through the aerated bars is sensitive to the bubble size distribution but it is important to understand the effect that other product factor may have on these results. This is to determine whether the airborne ultrasound system is actually sensitive enough to monitor bubble size distribution. The application of chocolate crumb was used to represent what may happen in a manufacturing environment and the oil layer was used to represent a different surface condition. The results of the B scans (Figure 8) showed that the topography of the chocolate products has a large affect on the attenuation with a variation of approximately 15 dB. The attenuation was greatest at the side of one of the domes in the bar, this is most likely due to the incident ultrasound wave contacting an angled surface and having a larger proportion reflected away from the receiver than transmitting through the bar. The area at the top of an individual domed section had the lowest attenuation even though this is the region with the most aerated chocolate that the ultrasound pulse must travel through. The crumb layer had very little effect on the overall attenuation (Figure 8 D) with only a small increase in attenuation in this region. The oil layers did cause an increase in attenuation but this was only small (Figure 8 E). The average attenuation was calculated from all points along the length scan and shown in Table 2. It can be seen here that the scans with the surface conditions had a slightly higher average attenuation but this was only approximately 3 dB.

Table 2. Average attenuation for different surface coatings

	Average attenuation (dB)	Standard deviation (dB)
None	72.3	0.5
Chocolate dust	75.2	1.2
Oil	75.4	1

The attenuation results for the aerated chocolate bars showed that at different frequencies the effect of the small and large bubble size distributions was between 3 and 12 dB. The topography of the bar affected the attenuation by a similar magnitude meaning that any online system would need to have an accurate way to know exactly where on the bar the acoustic measurement was made, and statistical techniques could also be used. Although this sounds challenging this is not as difficult as it appears. Possibilities include using a laser range finder device or using calibration and synchronization techniques with the current airborne ultrasound system. The results for varying surface conditions showed that these had a small effect on the attenuation but this was not as great as that caused by the topography or bubble size. These experiments actually had extreme versions of changes in surface properties so it is anticipated that the actual effect they would have on an industrial system would be less.

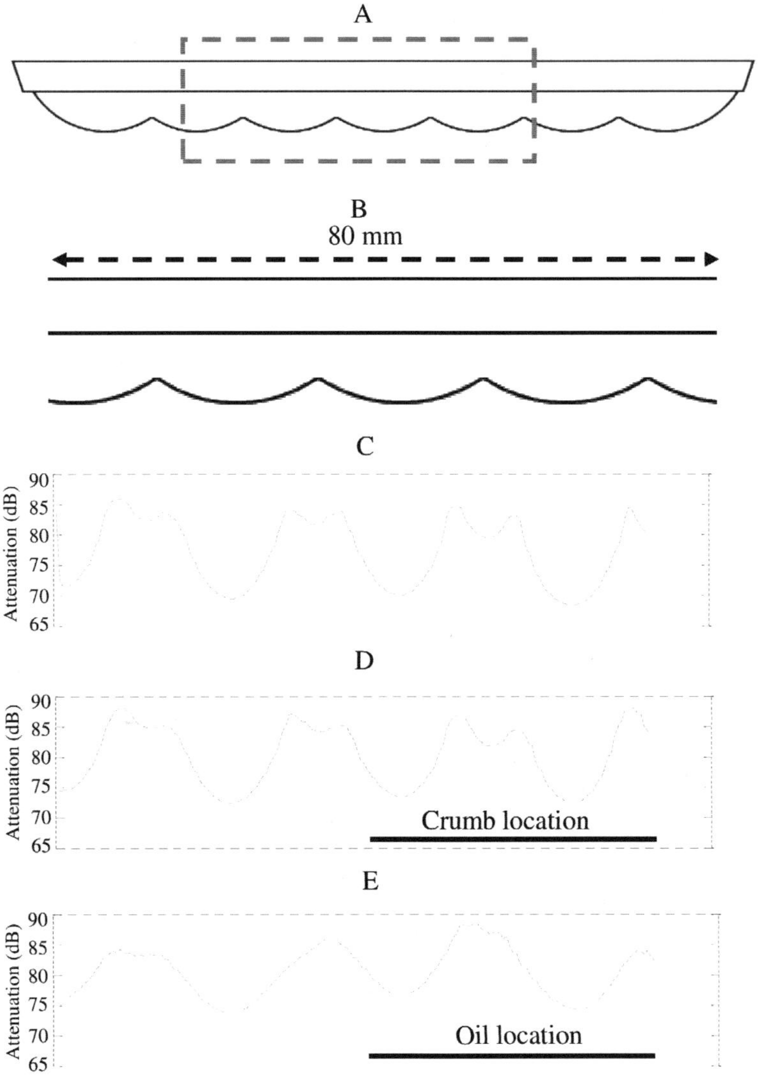

Figure 8. Airborne experiments and results along an aerated chocolate bar. A: the view of the side of the aerated bar. B: the actual length which was scanned. C: Attenuation results for an aerated bar as a function of length. D: Attenuation results for an aerated bar when chocolate crumbs were placed on the top and bottom of the bar at one end. E: Attenuation results for an aerated bar as a function of length when a thin oil layer was placed on the top and bottom of the bar at one end.

5. Conclusions

The purpose of this work was to determine the feasibility of using attenuation measurements by an airborne ultrasound system to monitor bubble sizes distribution in aerated chocolate bars. A computational model of the monitoring system and aerated chocolate was built to investigate whether the model could predict the experimental results and be used to optimise the design process. Experiments and simulations were performed on a 2D model system comprising of straws and aerated chocolate bars with a small and large bubble size distributions. The effect of the chocolate bars shape and surface properties on the ultrasound measurements was also investigated experimentally.

The results for the model two dimensional systems showed an increase in attenuation for the smaller straws and a good qualitative comparison between the experimental and simulation results. The results for the actual aerated chocolate bars also showed an increase in attenuation for the smaller bubble size. The simulations of the aerated bars did not show this affect. This was attributed to the model merely being a 2D representation of the experimental system whose image analysis may not fully represent the actual bubble size of the aerated chocolate product. The scans along the aerated bars showed that the shape of the bar had a great effect on the attenuation measurements but the surface condition (such as oil layer or chocolate crumbs) had little.

From this work it is concluded that airborne ultrasound could be used to monitor bubble size distribution in aerated chocolate products using attenuation measurements. More testing will be required to determine how accurately the system can detect small changes as the current work only studied two aerated products, one with a small and one with a large bubble size distribution. The results are encouraging as they indicate that the current technique would be capable of identifying an undesirable small or large bubble size distribution. The surface topography of the bars had a great effect on the attenuation measurements so the location of the measurement on the product must be known for meaningful interpretation of the results.

The computer simulations encouragingly recreated the experimental results for an identical geometry. With more precise modeling of the aerated chocolate products it will be possible to generate three dimensional models which could predict ultrasound propagation through the aerated bars and be used to aid the design process for an airborne ultrasound process monitoring system.

References:

[1] Junker B, Maciejak W, Darnell B, Lester M and Pollack M 2007 *Bioprocess Biosyst Eng.* **30** 313-326
[2] Povey M J W 1997 *Ultrasonic Techniques for Fluids Characterization, Academic Press*
[3] Dukhin A S and Goetz P J 2002 *Ultrasound for Characterizing Colloids, Elsevier*
[4] Strybulevych A, Leroy V, Scanlon M G and Page J H 2007 *Soft Matter* **3** 1388-1394
[5] Cobus L A E B, Ross K A, Scanlon M G and Page J H 2007 *j. Agric. Food. Chem.* **55** 8889-8895
[6] Strybulevych A, Leroy V, Shum A L, Koksel H F, Scanlon M G and Page J H 2012 *IOP Conf. Series: Material Science and Engineering* **42** 012037
[7] Elmehdi H M, Page J H and Scanlon M G 2003 *Journal of Cereal Science* **38** 33-43
[8] Pallav P, Hutchins D A and Gan T H 2009 *Ultrasonics* **49** 244-253
[9] Abramoff M D, Magalhaes P J and Ram S J 2004 *Biophotonics International* **11** 7 36-42.
[10] Mylavarapu P S K and Boddapati S 2011 *Proceedings of the 2011 COMSOL Conference in Bangalore*

Laser vibrometry characterisation of a microfluidic lab-on-a-chip device: a preliminary investigation

C Fury[1,2], P N Gélat[2,3], P H Jones[1] and G Memoli[2]
[1] Department of Physics and Astronomy, University College London, Gower Street, London, WC1E 6BT, UK
[2] National Physical Laboratory, Teddington, Middlesex, TW11 0LW, UK
[3] Department of Mechanical Engineering, University College London, Torrington Place, London, WC1E 7JE, UK

Since their original inception as ultrasound contrast agents, potential applications of microbubbles have evolved to encompass molecular imaging and targeted drug delivery. As these areas develop, so does the need to understand the mechanisms behind the interaction of microbubbles both with biological tissue and with other microbubbles. There is therefore a metrological requirement to develop a controlled environment in which to study these processes. Presented here is the design and characterisation of such a system, which consists of a microfluidic chip, specifically developed for manipulating microbubbles using both optical and acoustic trapping. A laser vibrometer is used to observe the coupling of acoustic energy into the chip from a piezoelectric transducer bonded to the surface. Measurement of the velocity of surface waves on the chip is investigated as a potential method for inferring the nature of the acoustic fields excited within the liquid medium of the device. Comparison of measured surface wavelengths with wave types suggests the observation of anti-symmetric Lamb or Love-Kirchhoff waves. Further visual confirmation of the acoustic fields through bubble aggregation highlights differences between the model and experimental results in predicting the position of acoustic pressure nodes in relation to excitation frequency.

1. Introduction

The biomedical application of microbubbles has been documented back to the 1960s [1]. But only in the last decades has the technology been developed to manufacture gas bubbles safe enough to inject into the human body. Bubbles with a radius in the range 2-8μm have a resonant volume oscillation excitation in reaction to MHz ultrasound, an effect which reradiates the impinging sound back out into the medium at a higher region of the frequency spectrum [2]. The resulting scattered field is highly responsive to the nature of the incoming ultrasound. As this frequency range is suited to the medical ultrasound scanner frequencies it has led to an increasing interest in microbubble use as biomedical sensors, contrast agents for ultrasound imaging and potentially as drug delivery vehicles [3]. As the sophistication of these applications increases, it is likely that the accurate modelling and understanding of how different bubble compositions respond to ultrasound will become increasingly important in ensuring the safety and efficacy of new products. There is therefore a requirement to develop a controlled environment in which to study the metrological aspects of the interactions between bubbles and sound fields.

Many studies have investigated the response of bubbles to sound fields. Early research into the acoustic emissions from cavitating bubbles [4] drew conclusions from large populations. It has been shown however that if bubbles are sufficiently close to one another they will interact [5] and this will in turn affect the emitted sound field. This indicates that it may be difficult to understand single bubble behaviour from experiments on bubble populations.

More recently, techniques have arisen to manipulate single bubbles to investigate their acoustic response in isolation [6][7]. These include techniques using kHz frequency acoustic fields which allow the position of bubbles in suspension to be controlled [7]. Using standing wave fields, the same technique may be used to trap bubbles at pressure nodes. The ability of this technique to manipulate many bubbles simultaneously makes it a potentially powerful tool for bubble characterisation by providing a facility which moves out of the large population regime.

The aim of this study is to characterise a custom lab-on-a-chip device designed for confining and measuring the response of microbubbles to MHz frequencies under controlled isolation. The microfluidic chip was designed and modelled using finite element modelling (FEM) at the National Physical Laboratory (NPL) and fabricated by Dolomite Microfluidics Ltd. We also discuss a scanning laser vibrometer method for characterising the acoustic fields excited within the chip. A similar technique has been employed by Neild *et al.*[8] in which a vibrometer is used to directly observe the vibration of a line upon a piezoelectric element in order to search for resonant excitation frequencies.

Here, this technique is extended to observe an entire driven surface near and far (up to 3 wavelengths) from a driving piezo element, to investigate standing waves in the excited structure that may suggest suitable frequencies for the acoustic trapping of particles. Comparison of the results with FEM allows verification and improvement of simulations.

The ability to accurately simulate what is happening inside the microfluidic channel will ultimately facilitate predictions of the pressure within the fluid. This cannot be measured directly experimentally as the introduction of a hydrophone with a size comparable to the channel would disrupt the acoustic field, significantly modifying the acoustic pressure values observed.

2. Microfluidic chips and experimental set up

The microfluidic chips used here have been designed to accommodate acoustic trapping, ultrasonic bubble excitation and optical tweezing. Simulations of resonant bubble motion [9] have shown that the bubbles will require a trapping force in the 1-10 nN range to maintain a stable trapped position. This magnitude of force cannot be produced by optical trapping alone and so must be complemented by acoustic trapping. The combined two tweezer system combines the relatively large acoustic trapping force with the high spatial resolution provided by optical tweezing.

The microchannel that contains the bubbles is labelled (1) in figure 1. The channel has four tubes that converge at the centre of the chip and are fed from the outer left and right edges. The straight portion at the bottom edge will be referred to as the 'main channel' and is the focus of the work described here. The diagonal channels are referred to as the left and right 'legs'. Pressure waves in the liquid contained in the channel are excited by a PZT (Lead zirconium titanate piezoelectric ceramic) transducer attached to the surface of the chip.

The microfluidic chip (maximum thickness 2.54 mm (in the central section containing the microchannel) and minimum thickness 1.2 mm (where the inlet ports are mounted)) was filled using a software controlled syringe pump (Kloehn, model 54022) and then sealed, so preventing the leakage of any fluid during use.

The cuboid in figure 1 (5.9 x 5.9 x 13 mm) is a PZT4 piezo transducer (Morgan Ceramics, UK) and its driving signal was generated with an Agilent 33250A signal generator providing 40mV_{pp} amplitude, amplified by an A&I 250L and impedance matched with a 1:25 toroidal transformer.

The laser vibrometer (Polytec Ltd., PSV-400) was used to scan the top surface of the microfluidic chip and record its velocity in the normal direction to the microchip surface during operation. The internal software synchronises the velocity data at each scanning point to the signal driving the PZT, obtaining the temporal evolution of the surface and its root mean squared velocity. In the set up used for this study, the device was triggered by the TTL output from the Agilent generator so that it is directly coupled to the PZT action.

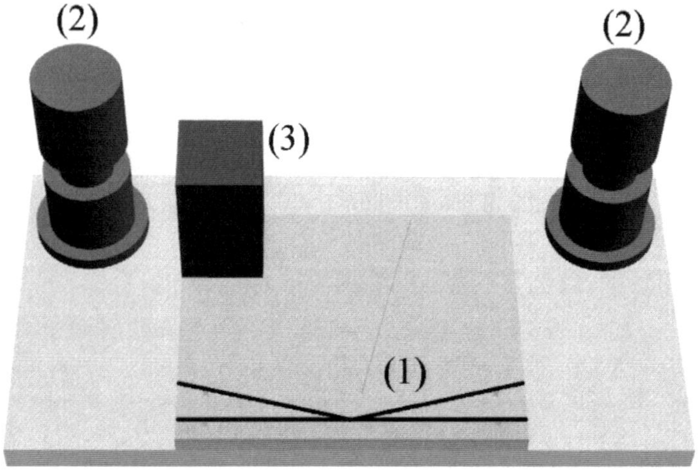

Figure 1. Microfluidic chip for optical and acoustic tweezing. The microchannel is visible at the lower edge of the chip in the form of a 'K' rotated 90° counter-clockwise (1). The cylindrical structures depict the fluidic in/out ports (2) and the cuboid is the PZT transducer used to drive the surface (3).

3. Modelling

In order gain a better understanding of the piezoelectric, mechanical and acoustical interactions between the driving transducer and the glass microfluidic chip, FEM simulations were carried out using the PAFEC VibroAcoustics software [10]. The model represents a 25 mm × 20 mm × 2.5 mm cuboid glass chip with water inlet holes and microchannels, together with a sinusoidally-driven cuboid piezo-ceramic transducer, as described in Section 2.

The mesh was generated using CATIA v5 Advanced Meshing Tools. Ten-noded tetrahedral elements with quadratic shape functions were used throughout the mesh, with a maximum separation of adjacent corner nodes of 1.5 mm in each element. For quadratic elements, it is well known that it is generally sufficient to mesh at three elements per wavelength. At 180 kHz, the employed mesh density corresponds to a lowest wave speed of 810 m s^{-1} indicating that the mesh density is likely to be higher than required for the materials used. The glass elements possess three translational degrees of freedom in the x, y and z direction of the global Cartesian axis set. The fluid elements have a single degree of freedom (acoustic pressure) and are coupled to the structural elements via continuity of the normal component of the particle velocity vector at the structure/fluid interfaces. The coupling of the tangential component of the velocity is neglected in this analysis.

Ambient temperature conditions of 25 °C were assumed. The physical properties of water were calculated using temperature dependencies obtained from the literature [11] and those for glass were initially taken from online databases [12][13]. These properties attained from databases are detailed in Table 1.

Table 1: Properties of glass and water.

	Density kg m^{-3}	Young's Modulus GPa	Longitudinal speed of sound m s^{-1}	Poisson's ratio
Glass [12][13]	2234	8.01	6694	0.27
Water [11]	997.9	Not applicable	1488	Not applicable

The piezoelectric elements were defined in terms of their full orthotropic, piezoelectric and dielectric properties, as defined in the PAFEC modelling system [10]. These elements possess four degrees of freedom: x, y and z translations in the global Cartesian axis set along with electrical potential.

The fluid boundary conditions at the inlets and outlets were assumed to be zero normal pressure gradients. Translations along the *z*-axis were restrained at the base of the chip. The electrical degrees of freedom were restrained across the bonded bottom electrode (earth electrode condition) and a sinusoidal electrical excitation of 1 V magnitude was applied at all nodes on the top electrode (live electrode). The frequency of excitation was varied between 130 kHz and 180 kHz in 0.5 kHz steps. At each frequency, the FEM calculations enabled the structural displacement vectors, acoustic pressures and electrical potentials to be obtained.

4. Excitation frequency determination

Two drawbacks of using the laser vibrometer were that the instrument could only be employed for a single frequency at a time and, with the present software, a frequency sweep cannot be automated. Therefore suitable frequencies needed to be chosen beforehand in order to efficiently investigate the regions of interest. The FEM data can be used to predict at which frequencies modes of the structure will occur by summing the pressure throughout the channel for each simulation frequency. Peaks in the spectrum of this "spatially averaged pressure" (figure 2a) will correspond to resonant modes of the channel and are therefore optimal candidates for acoustical manipulation experiments. The FEM model indicates these to be located at 135, 139, 172.5, 173 and 179 kHz.

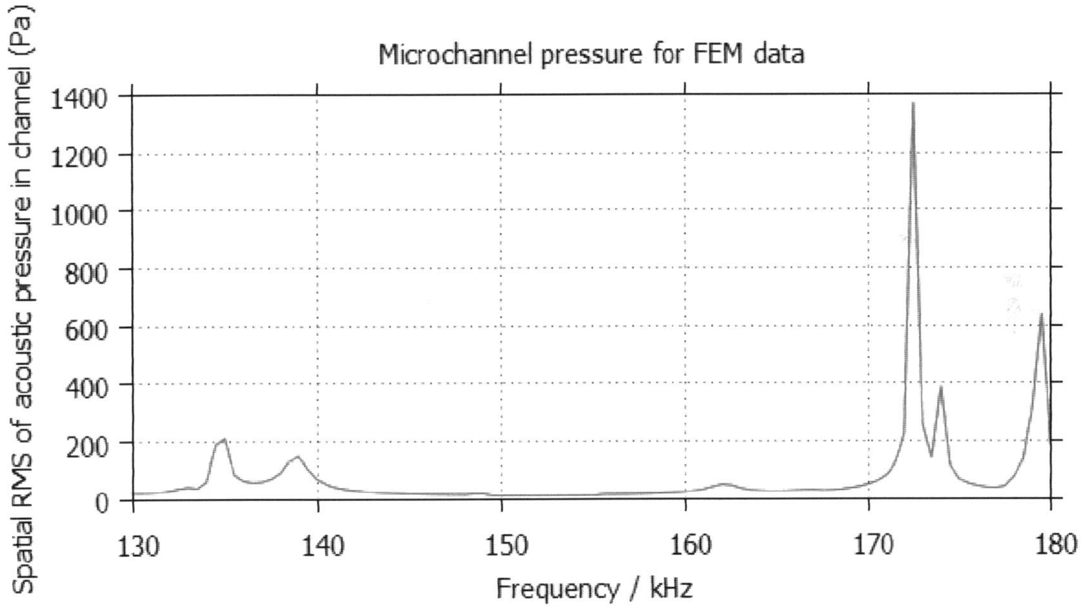

a)

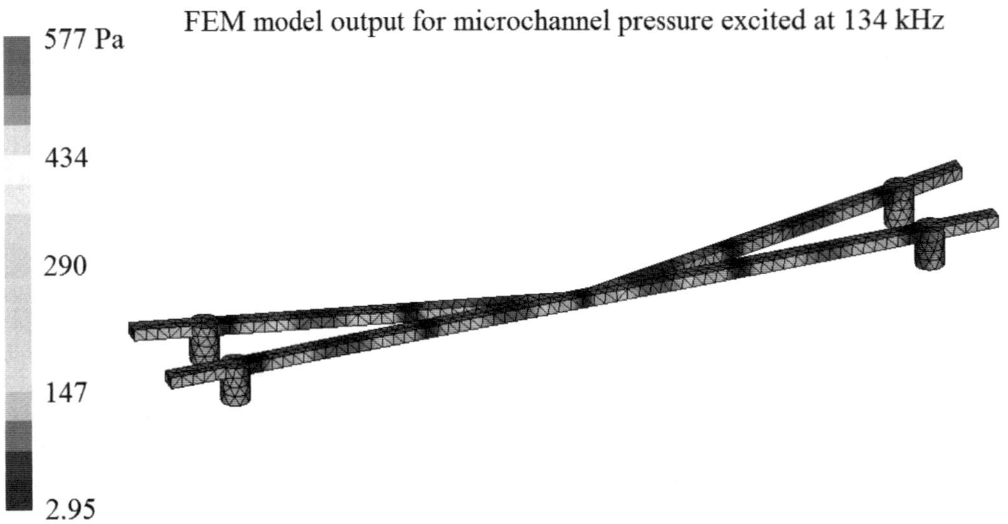

b)

Figure 2. Spatially averaged RMS of the acoustic pressure in the microchannel for frequencies in the range 130-180 kHz (in 0.5 kHz steps) derived from FEM simulations. Peaks denote regions where increased pressure is seen in the channel and are likely to indicate a trapping frequency. The inset model (b) shows a uniform standing wave pattern is predicted in the channel at 134 kHz.

The uncertainty as to whether the material properties reported in table 1 are relevant to this particular glass specimen and the simplifications in the FEM model are reflected in the ambiguity of the resonance positions in the spectrum of figure 2; while the order (and the relative position) of the peaks is potentially correct, their frequency may appear different in practice. To directly test the chip for excitation an electrical impedance scan was performed to investigate resonances [14]. The results can be seen in figure 3, where resonances appear as peaks in the Real Impedance data and from which two frequencies were decided upon for the preliminary tests described later within this paper: 104 kHz and 162 kHz.

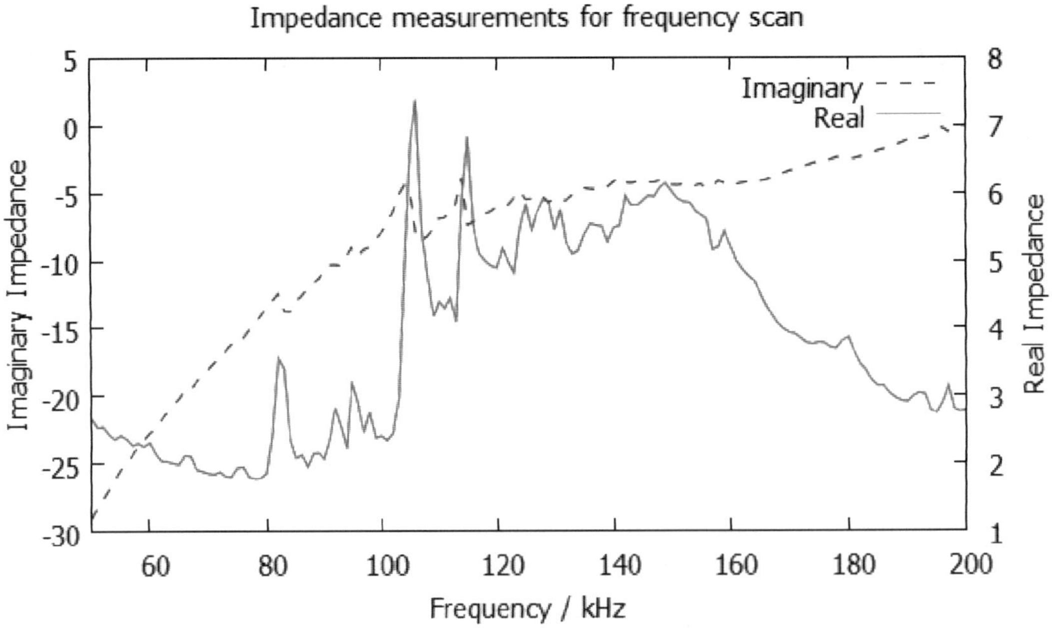

Figure 3. Electrical impedance data for the surface mounted transducer. Resonant frequencies occur for when there is a peak in the Real data and a negative slope in the Imaginary data.

5. Results

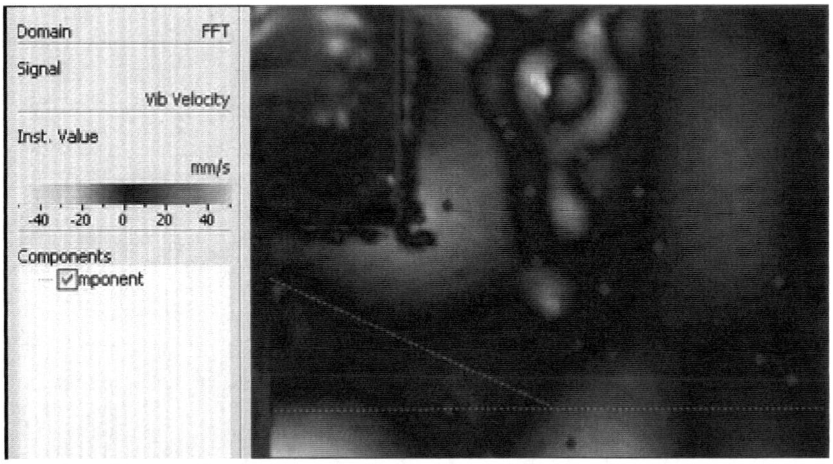

Figure 4. A sample of laser vibrometer data for a point in time during the driving frequency cycle at 162 kHz. The colour depiction denotes instantaneous velocity at each of the ~4500 data points in this scan.

Figure 4 shows an example data set from the vibrometer, with the colour scale representing the surface velocity at a point in time during the driving frequency cycle. The main channel has been identified as being located along the horizontal dashed line and the left leg along the diagonal line. Figure 5 shows a profile plot for the main channel at several evenly spaced phases in the driving oscillation cycle for 104 kHz. The position of the peak is denoted by the arrow and can be seen to be travelling to the right. For stable particle manipulation a standing wave must be observed and therefore this frequency is not a useful one. However, it should be noted that the magnitude of the velocity is high, likely suggesting that a structural mode is being excited.

Figure 6 shows similar information for a 162 kHz driving signal. Here the arrow marking the wave peak does not move in time, indicating a standing wave, this time with only an instantaneous velocity normal to the surface of one quarter of the magnitude of the 104k Hz case. A useful confirmation that the profile actually describes what is happening on the top of the microchannel can be seen in the data near the left-hand anti-node; there is a small region that appears to be oscillating 180° out of phase with the area around it. A post-scan inspection of the channel showed an air void within the liquid at this point. At a pressure free interface such as this we would expect a 180° phase change in the wave.

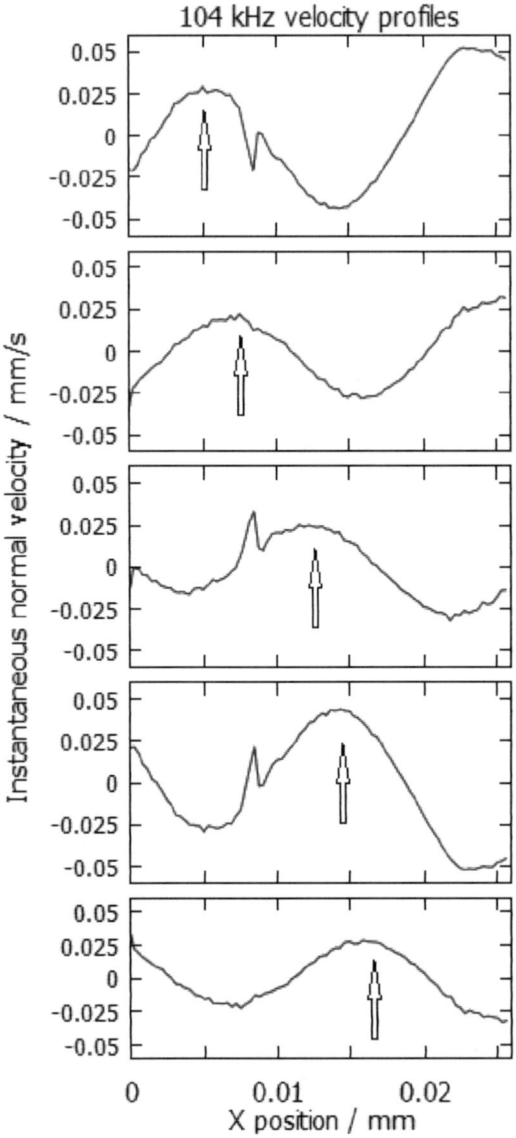

Figure 5. Normal velocity above the main microchannel for a 104 kHz excitation (vibrometer data). The wave observed is travelling to the right and will not provide suitable conditions for trapping. The arrow shows the movement of the wave peak.

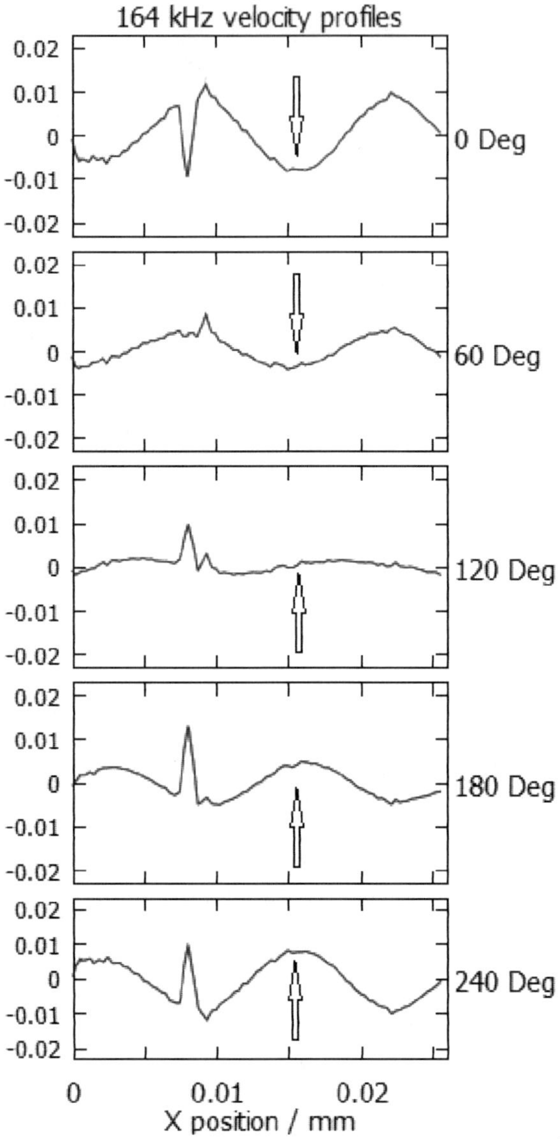

Figure 6. Normal velocity above the main microchannel for a 162 kHz excitation (vibrometer data). A standing wave can be seen with the peak denoted by the arrow. An air void shows a 180° phase change in wave velocity at the left hand antinode.

To prove that close to 162 kHz it is possible to aggregate particles, Expancel microspheres (AkzoNobel, 461-WU-40 polymer microspheres, 12-16 μm diameter) were pumped into the channel and a microscope used to capture their movement while the chip was excited through the PZT. After 20 seconds, grouped spheres could be observed (see figure 7), while no aggregation was observed at 104 kHz. Due to the size of the Expancel the clusters form at nodes in the field [15][16] and this can be used to measure the half wavelength of the standing wave. Figure 7 reports one of those values by the overlaid arrow, averaging between all visible nodal points gives a longitudinal wavelength in water of 8.9 ± 0.4 mm, which is comparable with the expected value of 9.3 mm (calculated using the speed of sound in water [11] and a driving frequency of 163.4 kHz). This node-node distance and the arrangement of nodes across the chip mirrors a state predicted by the FEM, given in figure 2b. The similarities are stark, but the two examples are performed at different frequencies with the experiment at 163.4 kHz and the model at 134 kHz. This raises a question over whether the model is complete enough (e.g. currently the damping terms are neglected) or if the tabulated data that was used to define the materials is incorrect.

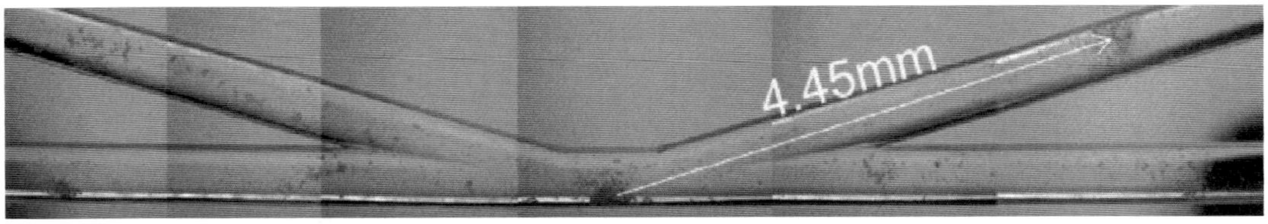

Figure 7. Composite image taken after aggregation of Expancel microspheres in a 163.4 kHz acoustic field. The distance between aggregation sites is λ/2 and the theoretical longitudinal wavelength in water agrees with the measured value of 8.9 mm. The Image was calibrated using a NPL graticule, with a resolution of 25 μm.

6. Discussion

In the previous section, we presented a comparison between wavelengths measured from the images of bubble aggregation and the ones predicted by the FEM model, finding that:

a. at 163.4 kHz, where aggregation and a standing wave have been observed, the pressure wave wavelength agrees with the theoretical one in water;
b. at 134 kHz the pressure wave wavelength predicted by the FEM agrees with that measured at 163.4 kHz, supporting a similar layout of the acoustic field; but not the theoretical wavelength in water at this frequency (11.2 mm).

As previously mentioned the reason for matching FEM modelling with experiments is related to the problem of measuring pressure in the microchannel. The laser vibrometer was able to determine whether a standing wave was present at the surface, but it was not clear whether it could it be used to infer the pressure distribution inside the channel.

Snell's law [17] predicts that at the interface between two different media (glass and water, in this case) the ratio of wavelengths is equal to the ratio of wave speeds in the two different media. Knowing the speed of sound and the measured wavelength in water, the corresponding values in glass and the type of wave that has been measured on the chip surface are needed.

The laser vibrometer was used to capture the wavelength of the surface wave at the top of the chip, giving a value of 13.3 ± 1 mm (figure 6), corresponding to a measured wave speed of 2154 ±160 m s^{-1} for a 162 kHz driving frequency. The first logical comparison is with the speed of shear waves, which can be obtained using the following equation [18]:

$$c_S = \left[\frac{E}{2(1+\nu)\rho}\right]^{1/2} \tag{1}$$

Where c_S is the shear wave speed, E the Young's modulus, ν is Poisson's ratio and ρ is the density.

Equation 1 and the values in table 1 give a shear wave speed of 3330 m s^{-1}, which is much higher than that measured. There were two likely reasons for this which are explored below.

The first was that the material data for the glass in table 1 was incorrect. Consequently the longitudinal wave speed in the glass was measured using a pulse-echo technique at 5600 ± 60 m s^{-1} (1 s.d.) and the density at 2594 ± 30 kg m^{-3} (1 s.d.); both values lie close to the tabulated ones. The Poisson ratio and the Young's Modulus of glass sit within a narrow band [15] and cause only a small change to the velocity result. Taking the mean values from tabulated data [12][13] (0.219 for the Poisson's ratio and 7.14x10^{10} Pa for the Young's modulus) produces a shear speed result of 3360 ms^{-1}, very similar to the one obtained from tabulated data and not significantly closer to the measured wave speed. Even if in future studies these unknown material properties will be measured, another wave description must be found in order to bridge the gap between experiment and current theory.

The second possibility is that the treatment of these waves as shear waves has been too simple and that a more involved wave model is needed. The waves we considered were Rayleigh, Lamb and Love-Kirchhoff waves, with the closest result being shared between the Lamb and Love-Kirchhoff regimes.

The Lamb wave is an elastic wave that establishes in a vibrating, free plate, of infinite planar size, and where the wave velocity is governed by a finite plate thickness. The thickness of the microchip being 2.5 mm, this would give a wave speed (Rose [19], P. 111 Equation 8.39) of 1757 m s^{-1}. The Love-Kirchhoff wave is similarly an elastic wave that propagates in a free plate but of both finite planar dimensions and thickness and governed by the 2D modes being excited. This gives an expected speed of 2299 m s^{-1}, using x = 4th mode y = 2nd mode as seen in figure 4. This is higher than observed by one standard deviation; closer than the Lamb wave which is lower by twice the standard deviation.

In particular, this study has only used 2 frequencies for the vibrometer, of which only one was found to be viable for trapping and comparisons. Future studies will include more frequencies and successfully pick appropriate trapping frequencies from the impedance data to compare against FEM results across the entire spectrum. Both the Lamb and Love-Kirchhoff waves are dispersive to different degrees. A wider investigation of excitation frequencies will show a tendency to favour one dispersion curve over the other and so deliver the correct wave type.

With a larger database, similarities or offsets in the real compared to FE-modelled frequencies may appear that can then be treated for by different effects such as damping. In the short term experimental work can be progressed by factoring out any offsets in the correct positions of the modelled frequencies to produce a more efficient trapping frequency predicting technique. Once a quantifiable shift or scaling of the frequencies can be found, we may begin to understand the causes of the disparity seen and update the model further.

7. Conclusions

This work presents the results of a preliminary study into the characterisation of surface waves upon a microfluidic chip that is intended for acoustic microbubble manipulation. A laser vibrometry technique was employed to measure surface normal velocities upon a chip excited at different frequencies. The technique was used successfully to detect standing waves upon the chip surface at a frequency that was also seen to generate a standing wave within the microfluidic channel inside of the chip; enabling the identification of frequencies suitable for acoustic trapping.

The technique was also shown to be able to detect air voids within the liquid, raising confidence in that it can be used to directly measure what is happening in the liquid even though it is imaging the glass surface above it. The ability to both observe the glass wave motion and infer the pressure distribution within the microchannel enable the data gained to be directly compared to finite element modelling of the chips.

So far the surface wave results have been unexpected and differ from modelling predictions. The surface velocities lay in a region between those expected for anti-symmetric Lamb waves and Love-Kirchhoff waves, with these being lower and higher respectively. Further work must be carried out to obtain the correct material properties for the glass so that the validity of the speeds from equation 1 and table 2 can be verified.

Additionally, the number of frequencies explored through the laser vibrometer will be increased to enable a more complete comparison with the FEM, underpinning an understanding of the frequency shift and allowing models to be adjusted. When the models produce a result comparable to observations they can begin to be used as a tool to interrogate the pressures involved in the acoustic manipulation, a measurement that is difficult to obtain in practice.

Acknowledgements

The authors acknowledge funding from EPSRC and the National Physical Laboratory Strategic Research programme. Useful discussions with C Harfield and E J Stride, University of Oxford, and B Zeqiri, NPL, are also gratefully acknowledged.

References

[1] Kremkau F W, Gramiak R, Shah P M, Carstensen E L and Kramer D H, 1970, "Ultrasonic detection of cavitation at catheter tips", *Am. J. of Roentgenology, Radium Therapy, and Nuclear Medicine*, 110 (1):177.

[2] Zeqiri B, Gelat, P N, Hodnett M and Lee N D, 2003, "A novel sensor for monitoring acoustic cavitation. Part I: Concept, theory, and prototype development," *Ultrasonics, Ferroelectrics and Frequency Control, IEEE Transactions on*, **50** (10), 1342-50.

[3] Blomley M J K, Cooke J C, Unger E C, Monaghan M J, and Cosgrove D O, 2001, "Microbubble contrast agents: a new era in ultrasound", *BMJ*, **322**.

[4] Neppiras E A, 1968, "Measurement of acoustic cavitation", *IEEE transactions on Sonics and Ultrasonics*,SU-**15** (2), 81-88.

[5] Garbin V, Ferrari E, Cojoc D, Fabrizio E, Overvelde M L J, Van der Meer S M, Versluis M, De Jong N and Lohse D, 2006, "Optical Trapping of Ultrasound Contrast Agent Microbubbles: Study of the Bubble-Wall and Bubble-Bubble Interaction in Ultrasound," *Ultrasonics Symposium 2006. IEEE*, 513-16.

[6] P Jones, E Stride and N Saffari, 2006, "Trapping and manipulation of microscopic bubbles with a scanning optical tweezer", *Appl. Phys. Lett.* **89**, 081113.

[7] Rabaud D, Thibault P, Raven J-P, Hugon O, Lacot E, and Marmottant P, 2011, "Manipulation of confined bubbles in a thin microchannel: Drag and acoustic bjerknes forces", *Physics of Fluids*, **23**.

[8] Neild A, Oberti S and Dual J, 2007, "Design, modeling and characterization of microfluidic devices for ultrasonic manipulation", *Sensors and Actuators B*, **121**, 452-61.

[9] Harfield C, Pawlikowska A, Stride E, Jones P H and Memoli G, 2011, "Opto-acoustic trapping of microbubbles", Poster presented at the *Microbubble symposium: Fabrication, characterisation and translational applications*, Leeds, UK.

[10] PACSYS, PAFEC tool, 2010, www.vibroacoustics.co.uk.

[11] Bilaniuk N and Wong G S K, 1993, "Speed of sound in pure water as a function of temperature", *J. Acoust. Soc. Am.*, **93**, 1609-12.

[12] Tables of Physical & Chemical Constants (16th edition 1995). 2.1.4 Hygrometry. Kaye & Laby Online. Version 1.0 (2005).

[13] Schott B270 glass technical sheet, http://www.pgo-online.com/intl/katalog/B270.html , last accessed 04/04/2013.

[14] Barsoukov E and Macdonald J R, 2005, "Impedance Spectroscopy Theory, Experiment and Applications", 2nd Edition, New jersey, USA (Wiley).

[15] Lenshof A, Magnusson C and Laurell T, 2012, "Acoustofluidics 8: Applications of acoustophoresis in continuous flow microsystems", *Lab Chip*, **12**, 1210.

[16] Leighton T G, 1997, "The Acoustic Bubble", London (Academic Press).

[17] Gedge M and Hill M, 2012, "Acoustofluidics 17: Theory and applications of surface acoustic wave devices for particle manipulation", *Lab Chip*, **12**, 2998–3007.

[18] Pierce A D, 1989, "Acoustics: An Introduction to Its Physical Principles and Applications", Woodbury, NY (Acoustical Society of America), P. 130, Eq. 3-6.1.

[19] Rose J L, 2004, "Ultrasonic Waves in Solid Media", (Cambridge University Press), p. 111.

Enzyme activity determination using ultrasound

M J Holmes[1], T Southworth, N G Watson and M J W Povey.
The School of Food Science and Nutrition. The University of Leeds,Woodhouse Lane, Leeds LS2 9JT, UK

E-mail: prcmjh@leeds.ac.uk

Abstract. Here are presented the results of a novel approach to the measurement of enzyme reaction rates in which ultrasound velocity measurement is used. Our results show enzyme activity is observable, in the acoustic context, and that furthermore this offers the potential to estimate the rate of reaction over different substrate concentrations and temperatures. Findings are corroborated with optical microscopy and rheological measurements. Ultrasound velocity measurement can be performed without the need for aliquot extraction and offers an efficient, non-invasive and dynamic method to monitor enzyme activity.

1. Introduction

Enzymes are generally globular proteins that are highly selective biological catalysts which accelerate both the rate and specificity of metabolic reactions. Highly specific, they only react with particular substances, known as the subtrate which in turn is transformed into product molecules. The reactions may occur in several ways but all lower the Gibbs free energy [1]. It is thought that approximately 25,000 enzymes exist and the most common way to classify enzymes relates to type of reaction they catalyse [2, 3, 4]. These reactions are important for various uses within the food, chemical, biofuel and medicinal industries. For example, in the food industry, a mixture of pectinases and amylases are used to extract and clarify fruit juice [5]. In this investigation α-amylase, and diastase (α-β-amylase) will be under consideration. Without the presence or addition of of enzymes, reactions would not only progress at a much slower rate, but some may not occur at all. The measurement or estimation of the rate of reaction, known as the enzyme kinetics, can be calculated using a variety of assays and techniques, for example spectrophotometric, fluorometric or chromatographic assays.

There are various factors governing enzyme reaction rates which must be considered when determing enzyme kinetics. Examples being pH, temperature, salt concentration [6] and the concentration of both enzyme and substrate. Typically an enzyme – substrate reaction can be expressed using the Michaelis – Menten equation where the maximum rate of reaction is measured, known as V_{max}, and the concentration of substrate at the midway point of the reaction is measured, known as K_m (often refered to as the Michaelis constant) [4, 7, 8]. Different forms of biochemical analysis are conducted in order to determine the Michaelis – Menten rate constants. Techniques used to determine reaction rates, such as spectrophotometry, may have certain issues that must be considered in order to accurately determine rate reactions. For example, Bisswanger (2004) describes the use of ultraviolet spectroscopy to detect light absorption of a substrate at a particular wavelength; however, absorption at the same wavelength for the product formed was also noted thus creating inaccuracies in the substrate-product kinetics [2].

[1] To whom any correspondence should be addressed.

Background knowledge of the nature of the enzyme and the assay being used is recommended, creating difficulties for those who are less experienced in the field. Analysis techniques can be time-consuming, expensive and invasive with the requirement of aliquots. Discontinuous assays, are labour intensive and due to a 'stop and sample' methodology they sometimes will provide an estimation of the kinetics rather than an accurate measurement [9]. In some cases, techniques such as titrimetric methods are described as not being sufficeintly sensitive enough for use, conversely use of radiochemical methods, although sensitive, are not desirable due to the handling of radioactive materials [10]. Additionally, the rate of reaction is typically calculated using the initial rate, rather than the overall reaction. This can create inaccuracies of its own, with some studies describing fluctuating catalytic rates [11] and issues relating to analysis of only the initial reaction rate [12].

This paper describes how using ultrasonic techniques can offer an alternative method to analyse enzyme kinetics whilst overcoming some of the issues with traditional methods mentioned above. Ultrasound offers a *real-time* method of analysis whereby the dynamics of both enzyme and substrate can be monitored removing the need to prepare and examine several aliquots. The process is non-invasive, non-destructive, may be conducted on optically opaque materials and measurements may automatically collected throughout the reaction minimising process and measurement uncertainties, for example in extraction and sample preparation required for further analyses.

By definition, ultrasonic waves have frequencies exceeding the upper limit of the normal human audible range, being approximately 16 kHz and consist of longitudinal pressure waves which oscillate particles in gases, liquids and solids in the direction parallel to the wave propagation and shear waves which correspond to shear deformations of the material. Ultrasound is now routinely used in many applications including material testing, medical imaging and particle sizing and enables characterisation of the mechanical properties of the medium, e.g. elastic moduli, density, viscosity and phase state [13, 14]. Ultrasonic techniques have become increasingly beneficial for the characterisation of compounds and has been demonstrated to discern molecular structures of compounds [15]. In particular, within the food industry the ability to classify foods [16], assess quality [17], and analyse the composition of food components during processing or over shelf life [18]. It is a technique that is non-invasive, non-destructive and can penetrate opaque materials. Consequently, detailed analysis of a variety of foods can be quickly collected, providing real – time information. A greater understanding of textural and rheological properties can be gained [19, 20], and in relation to this paper, the measurement of processes and reactions through change in viso-elastic density in starch samples are investigated as a novel approach.

Ultrasonic measurements typically consist of the speed of sound v [ms^{-1}] a wave will propagate through a medium and the associated attenuation α [dB MHz^{-1} cm^{-1}] in the incident wave intensity. Under the assumptions of negligible thermo-acoustic losses, the acoustic parameters may be related to the complex elastic modulus $M = M' + iM''$ of the medium where;

$$M' = \rho v^2 \tag{1}$$

$$M'' = \frac{2\rho\alpha v^3}{\omega} \tag{2}$$

Here, ρ is the sample density (kg m^{-3}) and ω the angular frequency (rads s^{-1}) of the wave. The relationships to the conventional rheological parameters of bulk and shear moduli K and G respectively (having the units of Pascals Pa), are as follows;

$$M = K + \frac{4}{3}G \tag{3}$$

where,

$$G = G' + iG'' = \frac{\sigma_0}{\varepsilon_0}(\cos\delta + i\sin\delta) \tag{4}$$

During rheological measurements an applied strain amplitude ε_o generates a stress response σ_o with phase lag δ between the stress-strain oscillations. In equation (4), G' is referred to as the storage modulus and G'' the loss modulus [13].

In hydrogels, the bulk moduli K is of the order GPa and is dominated by the water phase which has a shear modulus in the order 10^{-5} Pa and so any shear waves are evanescent. Primary contribution toward the shear modulus manifests through the gel network and is of the order kPa. On this basis and upon examination of equations (1), (3) and (4) one can verify that the speed of sound in liquids is dominated by the bulk modulus giving rise to the approximation commonly quoted in equation (5);

$$v = \sqrt{\frac{K}{\rho}} = \sqrt{\frac{1}{\kappa\rho}} \tag{5}$$

where κ is the bulk compressibility. In solids, since shear waves may be supported in the medium, the corresponding relationships are;

$$c_L = \sqrt{\frac{K + \frac{4}{3}G}{\rho}} \tag{6}$$

$$c_S = \sqrt{\frac{G}{\rho}} \tag{7}$$

where c_L is the associated longitudinal or bulk wave speed and c_S the shear wave speed and of the same order of magnitude with the latter typically reduced by a factor 2-4 [14]. In gels the contribution of the shear storage G' modulus toward the velocity is minimal but in the current ultrasonic context is measurable and offers the capacity to monitor the enzyme action upon the substrate gel network.

Ultrasound is commonly used to study the gelatinisation of starch [21, 22, 23]. In the presence of heat and water, starch undergoes a series of changes leading to gelatinisation. Intermolecular bonds are broken down, allowing hydrogen bonding sites to interact with water. Water is able to diffuse into the armorphous spaces within starch granules causing swelling. This swelling breaks internal structures and granules split. A three-dimensional network forms via starch cross-linking which causes the solution to thicken and form a gel. The temperature of gelatinisation is specific to the type of starch. Gelatinisation temperatures are on average between 50 °C to 80 °C, and in some cases modified starches exceed this and gelatinise at far higher temperatures. It is dependent on amylopectin cross linkages and amylose/amylopectin ratios. Decreased amylose content equates to an increase in gelatinisation temperature. According to Belitz *et al.* (2004) potato starch has an amylose content of approximately 23% and a gelatinsation temp between 58-66 °C. Waxy corn (commonly known as maize) starch has between 0-1% amylose content and a gelatinisation temp of 63-72 °C [24].

Enzyme activity will be measured using gelatinised starch solutions with the addition of α-amylase or diastase (α and β amylase derived from malt) to catalyse the breakdown of the starch. Measurements will be taken to determine the ultrasound velocity as the starch solution is metabolised by the enzyme.

Using this approach, it is possible to monitor the reaction between enzyme and substrate through change in speed of sound. This is an efficient, *real time* process allowing the enzymic activity to be observed rather than just the initial rate. The measurements reported here improve on work first carried out in 1983 by Povey and Rosenthal whom investigated the ultrasonic detection of the degradation of Starch by α-amylase [25].

In the following sections we shall describe the use of ultrasound to monitor enzyme activity upon low concentration gelatinised starch gel samples. To corroborate findings ultrasound measurements will be taken using two devices, the Cygnus UVM ultrasound velocity meter (Cygnus Instruments, Dorchester, U.K) and the Resoscan TF Instruments. Additionally, rheological measurements (Kinexus Rheometer, Malvern, UK) of the gels pre and post enzyme to illustrate the changes in gel network properties and optical microscopy (Celestron LCD Digital microscope Model No. 44340) to show starch cell breakdown.

2. Materials and methods

2.1. Sample preparation

Starch solutions of 3% w/v and 5% w/v were prepared by the addition of native potato starch (NPS) or maize starch (at either 3 g or 5 g) to 100 ml millipore water (0.22 μm filter) and heated at 70 °C for 30 minutes to to ensure the starch is gelatinised and all granules are fully hydrated. Samples were continuously stirred with a magnetic flea to ensure uniform heating and samples were thoroughly mixed. The presence of dehydrated starch granules will disrupt attenuation and cause inaccuracies. The sample is then cooled to the required temperature (typically 37 °C) for analysis to take place.

A 1% w/v enzyme solution was prepared by either 100 mg α-amylase or diastase in 10ml millipore water and heated to 37°C to ensure the enzyme was fully hydrated but not inactivated. α-amylase (MP Biomedicals, LLC) origin from Bacillus subtilis, containing approximately 165 bacterial Amylase units per mg. One unit will dextrinize 1 mg of starch per minute at pH 6.6 and 30°C. Diastase (α and β amylase) (Fisher Scientific, UK) was derived from malt.

2.2. Ultrasound measurements

Ultrasonic velocity measurements were taken using a Cygnus UVM ultrasound velocity meter (Cygnus Instruments, Dorchester, U.K) schematic shown in figure 1 and, for comparison using a Resoscan instrument (TF Instruments, Heidelberg, Germany [26]) see figure 2.

2.2.1. Cygnus UVM. The UVM comprises of a cylindrical steel vessel (interior diameter 40 mm, capacity 75 ml) equipped with a 2MHz ultrasonic transducer which emits pressure wave pulses which travel through the sample medium hits the cell wall and returns to the transducer where the time of flight and the temperature will be recorded simultaneously. The cell is located within temperature controlled tank with 0.1°C uniformity (Grant Instruments, model GE15). The sample was stirred continuously using a magnetic flea to ensure a consistent temperature throughout the sample. This aspect is important since ultrasound velocity is temperature dependent. For example, in water at 25 °C every 0.1 °C change in water equates to a 0.3m s^{-1} change in velocity [13]. The UVM is initialised using a certified referenced temperature probe (Fluke, CHUB-E4 1529) and associated reference velocity. The UVM takes measurements 10 times a second and records the average and standard deviation (coefficient of variation less than 0.01%). This process allows changes in state of the medium to be correlated with ultrasound velocity measurements.

Gelatinised starch solutions of 3% w/v or 5% w/v were examined by placing 100 ml into the UVM cell with the addition of 1 ml of 1% w/v enzyme aliqout (10 mg enzyme). Upon the addition of either α-amylase, β-amylase or diastase into the cell, catalysis of the polysaccarides in the viscous soultions cause a reduction in viscosity equating to an increase in the velocity of sound.

2.2.2. Velocity reading validation. To establish that the observed differences in velocity were not merely the addition of the enzyme addition, an aliqout of 1 ml of 1% w/v enzyme solution was pipetted into the UVM cell containing only millipore water at 37 °C and provided an approximate velocity of 1523.65 ± 0.1 ms^{-1}. After addition of the enzyme solution the recorded velocity was approximately 1523.65 ± 0.1 ms^{-1}. Since the UVM supplies average estimates, a T-test was applied with no statistically significant changes observed between pre and post-addition of the enzyme aliquot (p-value > 0.05).

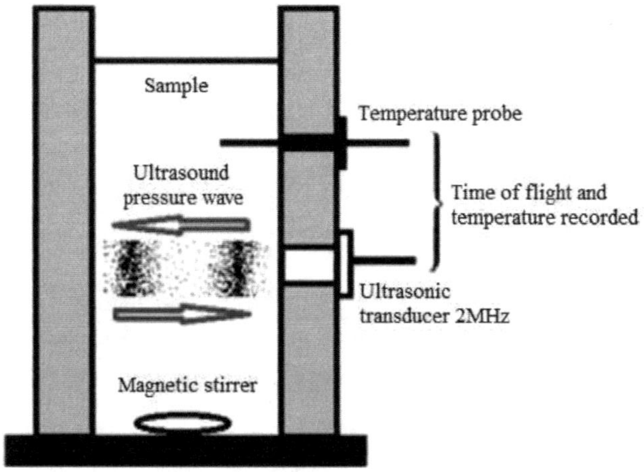

Figure 1. Schematic representation of the Cygnus UVM cell (Cygnus Instruments, Dorchester, U.K).

2.2.3.

2.2.4. Resoscan. Further ultrasound velocity measurements of gelatinised 3% w/v NPS solution were collected using a Resoscan, TF Instruments. This is a high precision relative measurement device (repeatability 0.01 ms^{-1}) which simultaneously collects data from two sample cells determining both the the ultrasonic velocity and attenuation of ultrasound pulses within the samples. Cells have a capacity of 200 µl, and the operational frequency range of 7.3-8.4 MHz, with the temperature controlled (±0.05 °C) by a Peltier thermostat. Data was automatically gathered by the instrument providing ultrasound velocity readings for starch solutions pre and post-enzyme addition.

Initially both cells contained 200 µl millipore water to verify accuracy against reported literature sources [13, 14]. Readings from both cells were consistent with one another. 100 ml gelatinised 3% w/v NPS solution was prepared, 200 µl was pipetted into each cell and readings were collected to ensure both cells were consistent with one another. The cells are flushed with millipore water and heated to 70 °C to remove residues before placing further samples into the cells.

Another batch of 100ml gelatinised 3% w/v NPS solution was prepared and kept on a heated plate at 37 °C containing a magnetic flea, 200 µl is pipetted into the first cell. Subsequently an aliqout of 1ml of 1% enzyme solution is pipetted into 100ml NPS sample which is stirred by the magnetic flea, then 200 µl is pipetted into the second cell.

It must be noted that there was an approximate 5 second delay between the addition of the enzyme to the solution and the placing of solution into the Resoscan cell.

2.3. Rheological measurements

Visco-elastic properties of the starch gels pre and post enzyme metabolised was conducted using a temperature-controlled Kinexus rheometer (Malvern, UK) using cone-plate arrangement (60 mm diameter cone with 2° angle and a 65 mm diameter plate). Approximately 2 ml of sample is applied to the plate at temperature is then changed to the required value 25 °C and, after 30 minutes of aging, measurements are performed. Oscillatory (0.1 – 10 Hz) rotational strains were applied to the sample, generating a vibratory stress to supply the complex shear moduli in the pre-enzyme gels as described in equation (3). Strain sweeps with shear rates 1-300 s^{-1} were implemented to generate apparent viscosities η (Pa s).

Figure 2. Resoscan instrument by TF Instruments, Heidelberg, Germany [26].

2.4 Optical microscopy

For qualitative analysis, optical microscopy was employed enabling still and video images of reactions between enzyme and substrate to be taken. All images have a 40X magnification. A Celestron LCD Digital microscope Model No. 44340 was used to visualise the time course of enzyme reaction taking place. Gelatinised starch samples of Native potato starch and maize starches were used with small quantities of enzyme solution were carefully pipetted onto the glass slide. Video sequences were captured enabling real time reactions to be visualised of the gelatinised samples before and after addition of the enzyme solution at increasing times upto 5 minutes.

3. Results and discussion

In the following sections various experimental results relating to ultrasonic detection of enzyme addition, velocity dependence upon temperature, starch sample concentration, rheology and optical microscopy will be presented.

3.1.Addition of enzyme in pure water and ungelatinsied samples.

Due to the sensitivity of ultrasonic techniques it was important to identify if the addition of the enzyme solution into the starch sample during analysis produced any significant changes in the speed of sound to verify that the observed change in ultrasonic velocity was not an simply an artifact of the addition of the enzyme solution or change in temperature. An aliqout of 1 ml of 1% w/v enzyme solution was pipetted into the UVM cell containing only millipore water at 37 °C and provided an approximate velocity of 1523.65 ± 0.1 ms^{-1}. After addition of the enzyme solution the recorded velocity was approximately 1523.65 ± 0.1 ms^{-1}. Since the UVM supplies average estimates, a T-test was applied with no statistically significant changes observed between pre and post-addition of the enzyme aliquot (p-value > 0.05).

This process was also applied to starch samples that had not been fully gelatinised (i.e. heated to temperatures < 50 °C, see figure 6 (a) and (e)) rendering the polysaccharides unvailable for metabolsim by the enzyme. No difference in ultrasound velocity was similarly noted.

3.2. Variation of ultrasound velocity measurements in relation to sample concentration.

Ultrasound measurements were collated from the UVM to examine how the concentration of the sample affects the speed of sound. Figure 3 shows 3% and 5% w/v gelatinised NPS solutions at 37°C. Velocity determinations were conducted using replicates. The data plotted in figure 3. Show running averages over 20 samples. As reported the UVM records average values based upon 10 measurements and also records the associated standard error (normally < 0.2 m s^{-1}) of the measurements. Note in Figure 3 that 3%w/v data have been translated by 4.5 ms^{-1} from 1526.5 ms^{-1} in order to show the overlay over the 5% w/v velocity data. α-amylase is added at approximately 200 seconds and subsequent hydrolysis with α-amylase of 3% and 5% w/v samples can be noted with an associated change in the ultrasound velocity. Enzyme activity rate can be seen to differ between the different concentrations. Enzyme action affects the speed of sound and the substrate concentration the difference in the final velocity achieved.

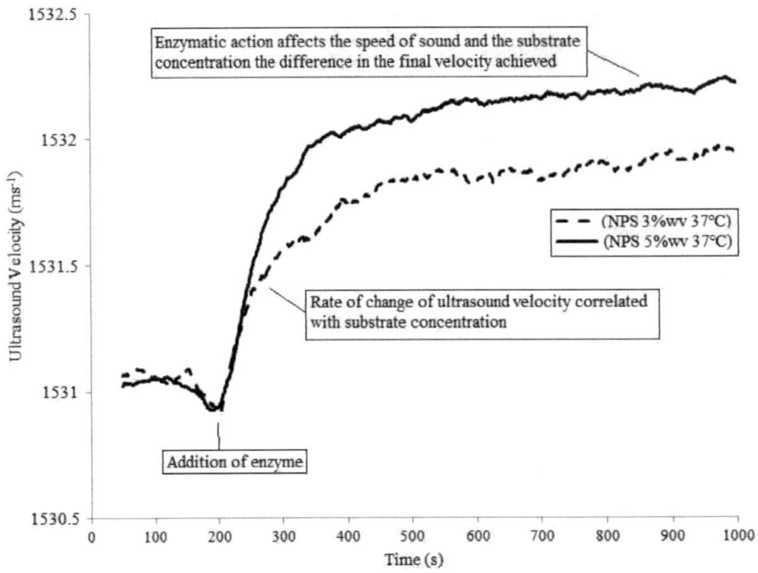

Figure 3. Ultrasound velocity (ms^{-1}) measurements against time (s) using the UVM for gelatinised Native Potato Starch (NPS) and water at 37°C; 5%w/v (solid line) and 3%w/v (dashed line)

The coefficient of variation is less that 0.01% therefore no errors bars are necessary on the plotted data. The initial velocities in the 3% w/v gel is 1526.6 ms^{-1} and in the 5% w/v gel 1531.0 ms^{-1} therefore the 3% w/v data has been translated by 4.5 ms^{-1} to enable comparison with the 5% w/v results. As can be seen, there is a increase in ultrasound velocity in both samples after the addition of the 1% enzyme solution at approximately 200 seconds (statistically significant p-values<0.001). Specifically, there is a greater absolute change in the velocity and in the rate also between the different concentrations. This demonstrates that ultrasound is able to detect enzyme rate activity over different concentration of the substrate and furthermore provides a real time observation of the kinetics thereby potentially removing the need for additional experimental analyses as discussed in the the introduction. The addtion of α-amylase results in the catalysis of starch from long chains, into short chained dextrins. This process in know as dextrinisation and can be seen in figure 3 to continue over approximately 6 minutes.

To ensure that the enzyme reaction had indeed metabolised all the substrate and that the enzyne itself had not been deactivated an additional aliquot of α-amylase solution was added to meatbolised sample. A further innoculation of 10mg α-amlyase showed no further change in velocity readings confirming that readings taken from the UVM relate to the reaction between enzyme and substrate itself rather than the addition of enzyme solution. This process was repeated with consistent findings.

3.3. Variation of ultrasound velocity measurements with temperature

Ultrasound measurements were collected using the UVM demonstrating changes in ultrasonic velocity for pre and post α-amylase addition in gelatinised 5% w/v NPS samples over several temperatures 26, 29 and 37 °C.

Figure 4 demonstrates that an increase in the speed of sound within the samples as the temperature is increased, consistent with an increase arising from the temperature dependent speed of sound in the water component of the starch gel. Furthermore the figure identifies an increase in ultrasonic velocity in post α-amylase addition. Post enzyme velocities reported are those when the velocity reaches a stable asymptotic value as noted in the inset graph.

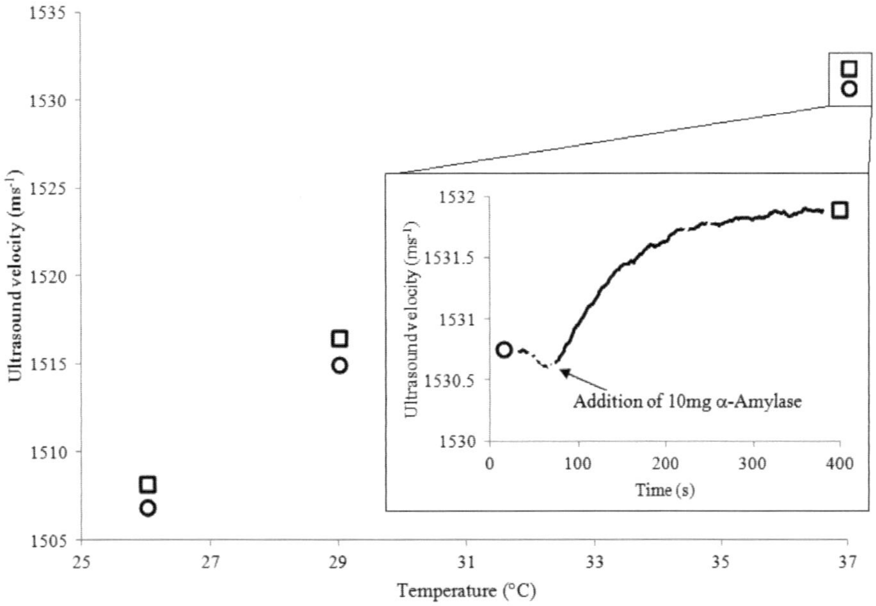

Figure 4. Speed of sound (ms^{-1}) in 5% w/v gelatinised Native Potato Starch (NPS) and water against temperature (°C). Circles show velocities prior to introduction of enzyme and squares after hydrolysis with α-amylase. Inset graph shows speed (ms^{-1}) against time (s) at 37 °C. The UVM performs ten velocity readings every second and the average with standard error (normally < 0.2 m s^{-1}) is recorded. Coefficient of variation is less than 0.01% so no error bars are shown.

3.4 Rescoscan measurements

Values taken from the UVM and Resoscan were compared to ensure similar relative pre and post enzyme velocity measurements were recorded. Table 1 shows speed of sound readings taken from both the UVM and the Resocan, giving equivalent readings for both pre and post α-amylase addition. This suggests that there is a significant level of repeatability when using these techniques.

Table 1. A table showing comparisons of measured ultrasound velocity ν (ms^{-1}) of NPS 3%w/v using the Resoscan and the UVM.

Instrument	Temperature/ precision (°C)	Velocity precision (ms^{-1})	Velocity (ms^{-1}) Pre α-amylase	Velocity (ms^{-1}) Post α- amylase
Resoscan[a]	37.0 ± 0.05	± 0.01	1526.7	1527.8
UVM[b]	37.0 ± 0.1	± 0.1	1526.7	1527.8

[a] Simultaneous velocity and attenuation taken in the frequency range 7.3-8.4 MHz.
[b] UVM frequency 2MHz, ten velocity readings taken every second and the average with standard deviation recorded. Coefficient of variation less than 0.01%.

Although the Resoscan is a high precision device it is more problematic in measuring velocity and attenuation in the gelatinised starch samples. This is primarily due to the small sample sizes required which makes it difficult to prepare the samples and also to monitor the evolution of the sample since premixing is required before presentation to the sample cells.

3.5 Rheology measurements

Rheological measurements of the 3% and 5% w/v gelatinised NPS starch gels pre and post enzyme metabolised were conducted using a temperature-controlled Kinexus rheometer (Malvern,UK) using cone-plate arrangement. Figure 5 shows the results obtained from the frequency sweep tests which show a reduction in apparent viscosity of 3% and 5% samples as shear rate increases. Furthermore, it identifies significant reduction of viscosity in samples post enzyme showing approximately 4-5 orders of magnitude reduction in the apparent viscosity η (Pa s). Measured stress measurements varied between 20-600 Pa in the gel phase and between 10^{-3} and 1.5 Pa in the post enzyme more liquid phase solutions. The oscillation tests produce values of around 10^2 Pa storage modulus G' measurements and order 10 Pa for the loss modulus G'' in the gel phase. Reductions in the viscosity and storage modulus evidently impacting upon the longitudinal velocity measurement as indicated in equation (6). In Figure 5 enzyme activity reduces the viscosity by orders of magnitude resulting with in a small increase in ultrasonic velocity within the medium. Within the gels, the storage modulus G' is of the order 10^2 Pa and loss modulus G'' is of the order 10 Pa. Rheology measurements were performed using the Kinexus rheometer (Malvern,UK)

3.6 Optical Microscopy

Qualitative results of the enzyme action upon the starch gels is shown in Figure 6 in which images taken from video sequences are generated by observing NPS samples before complete gelatinsisation after gelatinisation and then the sequence after the addition of the enzyme solution which progressively breaks down the starch cells, specifically by dextrinisation of the polysaccharides. Figure 6 (a) shows the ungelatinised NPS solution heated to approximately ~ 50 °C therefore complete gelatinsation was not achieved. Note cell striations characteristic of partially swollen starch granules. Figure 6 (b) shows the initial gelatinised starch gel (as described in the introduction) at the point where α-amylase added to commence catalysis. Note several starch granules still remain. Figure 6 (c) α-amylase progressively breaks down the starch cells over several minutes continually reducing viscosity. Figure 6 (d) Gelatinised cells broken down completely, starch granules remain. Figure 6 (e) ungelatinised maize starch (MS) cells heated ~ 50 °C. Note smaller cell size. Figure 6 (f) Diastase α–β amylase added to initial gelatinised sample. Figure 6 (g) Diastase breaks down starch cells. Figure 6 (h) All starch cells will finally be broken down. All images 40X magnification using a Celestron LCD Digital microscope Model No. 44340.

4. Conclusion

We have investigated the use of ultrasound to monitor enzyme activity in gelatinised starch samples. Various concentration and temperature conditions were examined. The results indicate that the rate of enzyme reaction of different concentrations of enzyme or substrate can be measured by ultrasound velocity. The technique permits time dependent measurement over a wide temperature range. Changes in ultrasound velocity in pre-dextrinised gels and subsequently dextrinised were monitored dynamically and related to concentration levels of the substrate (polysaccharide) remaining. Differences in velocity are also evident between different concentrations of the starch samples and at different temperatures. Findings are corroborated with optical microscopy and rheological measurements. Ultrasound utilises acoustic pressure waves to probe the material properties of the sample medium and can be performed without the need for aliquot extraction; this offers an efficient, non-invasive, non-destructive and dynamic method to monitor enzyme activity in potentially a wide range of applications.

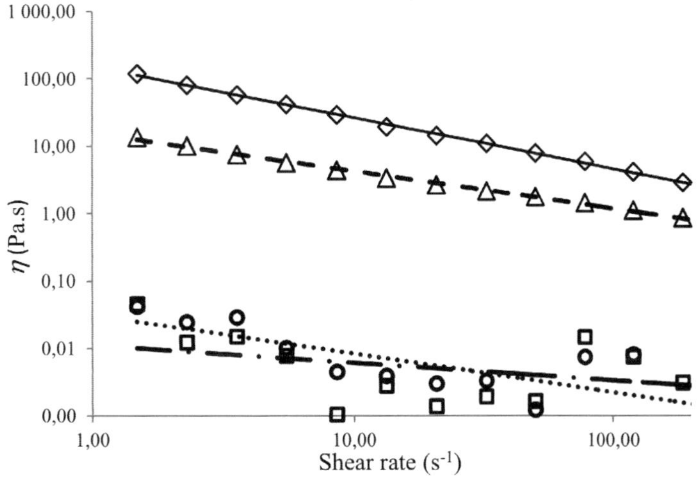

Figure 5. Apparent viscosity η (Pa s) against shear rate (s^{-1}) of gelatinised Native Potato Starch (NPS) and water at; 5% w/v (solid line with diamonds), 3% w/v (dashed line with triangles), 5% w/v after hydrolysis with α-amylase (dashed-dotted line with square) and 3% w/v after hydrolysis with α-amylase (dotted line with circles).

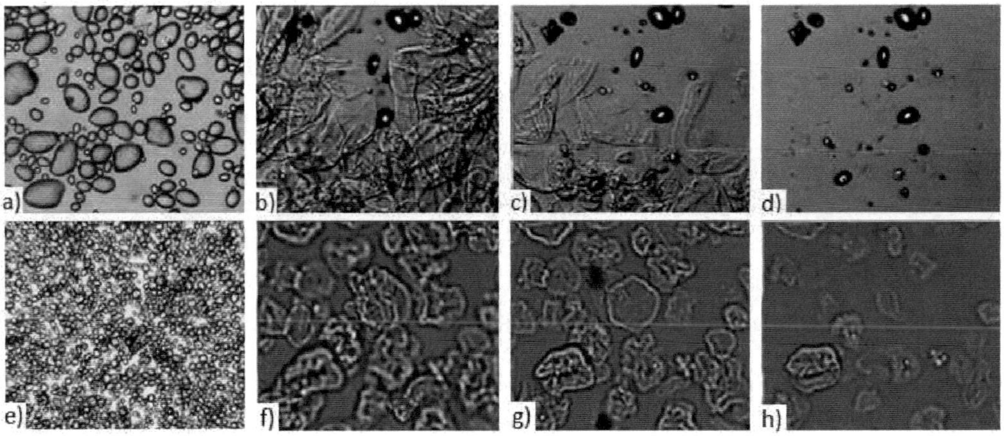

Figure 6. a) Ungelatinised NPS heated ~ 50 °C. Note cell striations characteristic of partially swollen starch granules. **b)** α-amylase added to gelatinised starch cells. Note starch granules. **c)** α-amylase breaks down starch cells over several minutes. **d)** Gelatinised cells broken down entirely, starch granules remain. **e)** Ungelatinised maize starch (MS). cells heated ~ 50 °C. Note smaller cell size. **f)** (Diastase) added α–β amylase. **g)** α–β amylase (Diastase) breaks down starch cells. **h)** All starch cells will finally be broken down. All images 40X magnification using a Celestron LCD Digital microscope Model No. 44340.

5. References

[1] Fersht, A. 1985. *Enzyme structure and mechanism*. (San Francisco: W.H. Freeman. pp. 50–2. ISBN 0-7167-1615-1).

[2] Bisswanger, H. 2004. *Practical Enzymology*. (Weinheim: Wiley).

[3] Voet, G. and Voet, J. 2011. *Biochemistry*. (4th Ed. Hoboken, N.J. Wiley).
[4] Roberts, M.B.V, Reiss, M.J and Monger, G. 2000. *Advanced Biology*. (UK: Nelson Thornes).
[5] Polaina, J. and McCabe, A.P. 2007. *Industrial Enzymes: Structure, Function and Applications*. (Spain: Springer).
[6] Anfinsen, C.B. 1992. *Advances in Protein Chemistry*. (San Diego: Academic Press)
[7] Johnson, K.A and Goody, R.S. 2011. The Original Michaelis Constant: Translation of the 1913 Michaelis–Menten Paper. *Biochemistry*. **50**, 8264-9.
[8] Cornish–Bowden, A. ed. 2004. *Fundamentals of Enzyme Kinetics*. (London: Portland Press Ltd).
[9] Lionettoa, F., Maffezzolia, A., Ottenhof, M.A., Farhat, I.A and Mitchell, J.R. 2006. Ultrasonic investigation of wheat starch retrogradation. *Journal of Food engineering*. **75** (issue 2), 258-66.
[10] Farooqui, A.A., Taylor, W.A., Pendley, C.E 2nd., Cox, J.W and Horrocks, L.A. Spectrophotometric determination of lipases, lysophospholipases, and phospholipases. *Journal of Lipid Research*. 1984, **25** (13), 1555-62.
[11] English, B.P, Min, W., Van-Oijen, A.M., Lee, K.T., Luo, G., Sun, H., Cherayil, B.J, Kou, S. C and Xie, X.S. 2005. Ever-fluctuating single enzyme molecules: Michaelis-Menten equation revisited. *Nature Chemical Biology*. **2**, 87-94.
[12] Schnell, S and Maini, P.K. 2003. A Century of Enzyme Kinetics: Reliability of the K_m and V_{max} Estimates. *Theoretical Biology*. **8**, 169-87.
[13] Povey M.J.W. 1997. *Ultrasonic Techniques For Fluids Characterization*. (San Diego: Academic Press).
[14] Kaye, G. W. C., & Laby, T. H. (1995). *Tables of physical and chemical constants* (16th ed.). Essex, England ; New York: Longman.
[15] Povey, M. J. W., Hindle, S., Kennedy, J. D., Stec, Z. and Taylor, R. G. 2003. The Molecular Basis for Sound Velocity in n-Alkanes, 1-Alcohols and Dimethylsiloxanes. *Physical Chemistry Chemical Physics* **5**, 73-8.
[16] McClements, D.J. 1988. *The use of ultrasonics for characterising fats and emulsions*. PhD thesis, University of Leeds.
[17] Alava, J.M., Sahi, S.S., Garcia-Alvarez, J., Turo, A., Chavez, J.A., Garcia, M.J and Salazar, J. 2007. Use of ultrasound for the determination of flour quality. *Ultrasonics*. **46** (3), 270-6.
[18] Awad, T.S., Moharram, H.A., Shaltout, O.E., Asker, D and Youssef, M.M. 2012. Applications of ultrasound in analysis, processing and quality control of food: A review. *Food Research International*. **48** (2), 410–27.
[19] Aparicio, C., Reza, C., Elvira, P., Molina-Garcia, L., Martino, A.D and Sanz, P.D. 2009. Assessment of starch gelatinization by ultrasonic and calorimetric techniques. *Journal of Food Engineering*. 94 (3-4), 295-9.
[20] Zhou, B., Wang, Z.X., Zhao, Y., Brautigan, D.L and Zhang, Z.Y. 2002. The Specificity of Extracellular Signal-regulated Kinase 2 Dephosphorylation by Protein Phosphatases. *Journal of Biological Chemistry*. 277, 31818–25.
[21] Johnson, D. L. 1982. Elastodynamics ff Gels. *Journal of Chemical Physics* **77(3)**, 153-9.
[22] Ammann, J.J., Galaz, B., 2003. Sound velocity determination in gel-based emulsions. *Ultrasonics*. 41 (7), 569–79.
[23] Ammann, J.J., Apablaza, V., Galaz, B., Flores, C., 2005. Ultrasonic sound velocity measurement in samples of soft materials through under-resonance excitation. *Ultrasound in Medicine and Biology*. 31, 485–91.
[24] Belitz, H.D., Grosch, W and Schieberle, P. 2004. *Food Chemistry*. (Spain: Springer).
[25] Povey, M. J. W., and Rosenthal, A. J. 1983, Ultrasonic Detection of the Degradation of Starch by α-Amylase, *J. Food Tech*. **19**, 115-9.
[26] TF instruments Inc. *Products*. [online]. http://www.tf-instruments.com/

Acknowledgments

The authors would like to acknowledge valuable discussions with Dr Caroline Orfila in preparation of this work.

HIFU scattering by the ribs: constrained optimisation with a complex surface impedance boundary condition

P Gélat[1,2,†], G ter Haar[3] and N Saffari[2]

[1] National Physical Laboratory, Hampton Road, Teddington TW11 0LW, United Kingdom.
[2] Department of Mechanical Engineering, University College London, Torrington Place, London WC1E 7JE, United Kingdom.
[3] Therapeutic Ultrasound Group, Joint Physics Department, Institute of Cancer Research, Sutton SM2 5NG, United Kingdom.

[†] E-mail: Pierre.Gelat@npl.co.uk

Abstract. High intensity focused ultrasound (HIFU) enables highly localised, non-invasive tissue ablation and its efficacy has been demonstrated in the treatment of a range of cancers, including those of the kidney, prostate and breast. HIFU offers the ability to treat deep-seated tumours locally, and potentially bears fewer side effects than more established treatment modalities such as resection, chemotherapy and ionising radiation. There remains however a number of significant challenges which currently hinder its widespread clinical application. One of these challenges is the need to transmit sufficient energy through the ribcage to ablate tissue at the required foci whilst minimising the formation of side lobes and sparing healthy tissue. Ribs both absorb and reflect ultrasound strongly. This sometimes results in overheating of bone and overlying tissue during treatment, leading to skin burns. Successful treatment of a patient with tumours in the upper abdomen therefore requires a thorough understanding of the way acoustic and thermal energy is deposited. Previously, a boundary element (BE) approach based on a Generalised Minimal Residual (GMRES) implementation of the Burton-Miller formulation was developed to predict the field of a multi-element HIFU array scattered by human ribs, the topology of which was obtained from CT scan data [1]. Dissipative mechanisms inside the propagating medium have since been implemented, together with a complex surface impedance condition at the surface of the ribs. A reformulation of the boundary element equations as a constrained optimisation problem was carried out to determine the complex surface velocities of a multi-element HIFU array which generated the acoustic pressure field that best fitted a required acoustic pressure distribution in a least-squares sense. This was done whilst ensuring that an acoustic dose rate parameter at the surface of the ribs was kept below a specified threshold. The methodology was tested at an excitation frequency of 1 MHz on a spherical multi-element array in the presence of anatomical ribs.

1. Background

High intensity focused ultrasound (HIFU) enables highly localised, non-invasive tissue ablation and its efficacy has been demonstrated for treating a range of cancers, including abdominal tumours [2], [3].

One of the challenges of trans-rib high-intensity focused ultrasound treatment is the need to transmit sufficient energy through the ribcage to ablate tissue whilst minimising the formation of side lobes, and sparing healthy tissue. Ribs strongly absorb and reflect ultrasound. This may result in overheating of bone and overlying tissue during treatment, leading to skin burns [3], [4], [5]. Hence, care must be taken so that sufficient energy is delivered through the rib cage ensuring that acoustic pressures at the treatment location are above the ablation threshold at the focus while at the same time maintaining the formation of side lobes to a minimal level. Approaches using multi-element piezoelectric transducer arrays in conjunction with multi-channel electronics have been suggested, where the magnitude and phase of each transducer element may be adjusted to help minimise acoustic pressures on the ribs. Commonly, pre-operative treatment planning strategies using such phased arrays rely on the deactivation of transducer elements obstructed by ribs, or so-called binarised apodisation [6], [7], [8]. Whilst this approach is convenient in its implementation, deactivation of elements will significantly reduce the ultrasonic power delivered at the focus and an increase in treatment time may occur as a result. Furthermore, it does not attempt to address the inverse problem of optimising the magnitudes and phases of the transducer element drive voltage in order to induce tissue necrosis at the required foci, whilst keeping the pressure on the ribs below a chosen threshold, and ensuring minimal formation of side-lobes.

Some of the limitations of binarised apodisation can be addressed in an approach based on the decomposition of the time reversal operator (DORT) [9]. The application of the DORT method to focusing through the ribs has been discussed by Cochard *et al* in 3D [10]. Using a singular value decomposition of the inter-element array response matrix, an excitation weight vector is obtained which is orthogonal to the subspace of emissions focusing on the rib. When applied to the array, this excitation vector enhances the acoustic energies deposited at the focal point compared with those on the ribs. Furthermore, Ballard *et al* proposed an experimentally validated method of an adaptive, image-based refocusing algorithm of dual-mode ultrasound arrays in the presence of scatterers [11]. This approach and the DORT method both have the advantage of not requiring *a priori* knowledge of the location of the ribs. Nevertheless, neither method directly quantifies energy depositions on the surface of the ribs, both relying instead on information from the back-scattered signal.

The inverse problem of focusing the field of a multi-element HIFU array inside the rib cage whilst ensuring that the pressure magnitude on the surface of the ribs does not exceed a given damage threshold, was addressed by Gélat *et al* [12]. The forward model employed a boundary element approach in which it was possible to address the effects of scattering and diffraction on 3D anatomical data [1]. In both these studies, the ribs modelled as perfectly rigid scatterers: this simplification is somewhat limiting as, although rib bone does indeed possess much higher impedance than soft tissue, ribs are not perfect reflectors.

Although there is no universally recognized meaning 'acoustic dose', in the context of this paper acoustic dose rate is defined as the time-averaged rate of heat generation by relaxational absorption per unit volume, following Duck [13]. Under continuous wave conditions and for linear wave propagation, this quantity $\langle q_v \rangle$ may be expressed as follows [14],

$$\langle q_v \rangle = \alpha \frac{|p|^2}{\rho c} \tag{1}$$

where

- α is the absorption coefficient arising from relaxation mechanisms when the shear viscosity is set to zero
- $|p|$ is the acoustic pressure magnitude
- ρ is the medium density
- c is the velocity of sound

The brackets $\langle \ \rangle$ denote a temporal average. These will be omitted in the remainder of this paper and the acoustic dose rate will be referred to as q_v.

In this study, the boundary element approach described in [1] and [12] was extended to incorporate an acoustic impedance boundary condition on the surface of the scatterer. This implementation was validated against an analytical solution for the scattering of a plane wave by a non-rigid sphere. Calculations using the forward model were then carried out on a rib mesh obtained from anatomical data. The acoustic source was 256-element spherical array with a pseudo-random distribution of elements on its surface. The inverse problem was solved using the same approach as in [12]. Using a Numerical Algorithms Group (NAG) Numerical Library routine [15], which calculates the minimum of a sum of squares for a set of nonlinear constraints, optimal values for the real and imaginary parts of the element velocities were obtained so that the acoustic pressures at specified locations in the exterior domain best fitted a required field distribution in a least-squares sense. Implementation of the constraints ensured that the acoustic dose rate on the surface of the ribs did not exceed a specified threshold and that the element velocity magnitudes were kept within the specified dynamic range.

2. Theory

Consider a scatterer defined by a closed smooth surface *S*. Let the medium surrounding the scatterer be defined as the exterior domain V_{ext}. Let the boundary condition at infinity satisfy Sommerfeld's acoustic radiation condition. The propagation of time-harmonic acoustic waves in a homogeneous isotropic inviscid medium is described by the Helmholtz equation:

$$\nabla^2 p(\vec{r}) + k^{*2} p(\vec{r}) = 0, \forall \vec{r} \in V_{ext} \tag{2}$$

where p is the acoustic pressure, $k^* = \omega / c^*$ is the acoustic wave number where ω is the angular frequency, c^* the complex sound speed in the external medium and $\vec{r}$ is the position vector. A source term may be included in the right hand side of (2) (incident pressure wave), but it is convenient to split the total pressure $p(\vec{r})$ as the sum of the incident pressure $p_i(\vec{r})$ and the pressure scattered by the surface *S*, $\forall \vec{r} \in V_{ext}$.

For scattering problems, the integral representation of the solution to (2) is given by the Helmholtz integral equation [16]:

$$\int_S \left[p(\vec{r}_q) \frac{\partial G(\vec{r}|\vec{r}_q)}{\partial n_q} - \frac{\partial p(\vec{r}_q)}{\partial n_q} G(\vec{r}|\vec{r}_q) \right] dS = \frac{1}{2} p(\vec{r}) - p_i(\vec{r}), \forall \vec{r} \in S \tag{3}$$

$$\int_S \left[p(\vec{r}_q) \frac{\partial G(\vec{r}|\vec{r}_q)}{\partial n_q} - \frac{\partial p(\vec{r}_q)}{\partial n_q} G(\vec{r}|\vec{r}_q) \right] dS = p(\vec{r}) - p_i(\vec{r}), \forall \vec{r} \in V_{ext} \tag{4}$$

where the Green's function in a three-dimensional space is given by:

$$G(\vec{r}|\vec{r}_q) = \frac{1}{4\pi\|\vec{r}-\vec{r}_q\|} e^{-ik^*\|\vec{r}-\vec{r}_q\|} \tag{5}$$

and $p(\vec{r}_q)$ is the pressure on the surface *S* at point $\vec{r}_q$ and $\vec{n}_q$ is the outward normal.

By using the linearised conservation of momentum equation on *S*, and by noting that ratio of the acoustic pressure and the normal component of the particle velocity vector on *S* is equal to the characteristic acoustic impedance of the scatterer, equations (3) and (4) may be rewritten as follows.

$$\int_S \left[p(\vec{r}_q) \frac{\partial G(\vec{r}|\vec{r}_q)}{\partial n_q} + ik^* \frac{\rho c^*}{\rho_1 c_1} p(\vec{r}_q) G(\vec{r}|\vec{r}_q) \right] dS = \frac{1}{2} p(\vec{r}) - p_i(\vec{r}), \forall \vec{r} \in S \tag{6}$$

$$\int_S \left[p(\vec{r}_q) \frac{\partial G(\vec{r}|\vec{r}_q)}{\partial n_q} + ik^* \frac{\rho c^*}{\rho_1 c_1} p(\vec{r}_q) G(\vec{r}|\vec{r}_q) \right] dS = p(\vec{r}) - p_i(\vec{r}), \forall \vec{r} \in V_{ext} \tag{7}$$

where

- ρ is the density of the acoustic medium in the exterior domain
- ρ_1 is the density of the scatterer
- c_1 is the speed of sound associated with the scatterer.

It is well known that the problem described by the discretisation of integral equation (6) suffers from non-uniqueness at frequencies of excitation approaching an eigenvalue of one of the (fictitious) modes of the cavity inside the scatterer. After discretising equation (6), the resulting matrix obtained becomes close to singular. For a wave number whose imaginary part is non-zero, solving the discretised form of equation (6) is often be sufficient, as the fictitious modes of the cavity inside the scatterer are likely to be damped out [17]. For the complex sound speed, frequency of excitation and scatterer dimensions considered in this paper, it has shown that a discretisation of the Helmholtz integral equation is indeed sufficient and that the coupling term involving the normal derivative formulation need not be considered [12]. This validation will was repeated as part of this work for a non-rigid spherical scatterer with a complex surface impedance condition. The results are shown in Section 3.

The formulation of the inverse problem is described fully in [12]. It is summarised below. By discretising equation (6), for all position vectors $\vec{r}$ on S corresponding to each node on the mesh of the surface, a linear system of equations may be generated.

$$[H]\{p_{surf}\} = -\{p_i\} \tag{8}$$

where $[H]$ is the boundary element matrix and the acoustic pressures on the surface S have been relabelled $\{p_{surf}\}$. It should be noted that the $[H]$ matrix referred to in this work differs from that in [12] in that it contains the contribution of terms associated with the integration of the normal derivative of the acoustic pressure multiplied by the Green's function over the surface S.

Let the incident acoustic pressure field be a linear combination of plane circular pistons rigidly vibrating in an infinite baffle. The incident field on S may be written as a linear combination of the source velocities of each piston as follows:

$$[H]\{p_{surf}\} = -[\beta]\{U\} \tag{9}$$

where the elements of $[\beta]$ may be obtained analytically in the far-field or by solving the Rayleigh integral in the near-field. $[\beta]$ is of dimension $M\times N$, where M is the number of nodes on the surface S after its discretisation and N is the number of plane pistons. $\{U\}$ is the vector of source velocities and is of dimension $N\times 1$.

Consider equation (6), where the position vectors are located in the exterior volume V_{ext}. For specified number of locations in the exterior volume, we have [12]:

$$[Q]\{p_{surf}\} = \{p_{ext}\} - [\gamma]\{U\} \tag{10}$$

where $[Q]$ is a matrix of weighting coefficients and the acoustic pressures in the exterior volume V_{ext} have been relabelled $\{p_{ext}\}$.

The vector of surface pressures may be eliminated by combining equations (9) and (10).

$$([\gamma] - [Q][H]^{-1}[\beta])\{U\} = \{p_{ext}\} \tag{11}$$

Equation (9) may be re-written as follows.

$$-[H]^{-1}[\beta]\{U\} = \{p_{surf}\} \tag{12}$$

or

$$[A]\{U\} = \{p_{surf}\} \tag{13}$$

where $[A] = -[H]\ ^{1}[\beta]\{U\}$.

Equations (11) may be re-written as follows.

$$[C]\{U\} = \{p_{ext}\} \tag{14}$$

where $[C] = [\gamma] - [Q][H]^{-1}[\beta]$.

When investigating an inverse problem involving complex quantities, it is convenient to reformulate in terms of purely real variables by rewriting equation (12) as follows:

$$\begin{bmatrix} \mathrm{Re}[C] & -\mathrm{Im}[C] \\ \mathrm{Im}[C] & \mathrm{Re}[C] \end{bmatrix} \begin{Bmatrix} \mathrm{Re}\{U\} \\ \mathrm{Im}\{U\} \end{Bmatrix} = \begin{Bmatrix} \mathrm{Re}\{p_{ext}\} \\ \mathrm{Im}\{p_{ext}\} \end{Bmatrix} \tag{15}$$

or

$$[\hat{C}]\{\hat{U}\} = \{\hat{p}_{ext}\} \tag{16}$$

We wish to obtain a set of real and imaginary parts of source velocities which best fit a prescribed field pressure distribution in a least-squares sense such that:

- the acoustic dose rate on S as defined in equation (1) does not exceed a threshold defined by $q_{v_{max}}$
- the source velocity magnitudes do not exceed the upper bound of the dynamic range of each element U_{max}.

This may be expressed as a least-squares minimisation problem with nonlinear constraints:

$$\min_{\hat{U}} \frac{1}{2} \left\| [\hat{C}]\{\hat{U}\} - \{\hat{p}_{ext}\} \right\|_2^2 \text{ such that } \begin{cases} q_v \leq q_{v_{max}} \\ |U| \leq U_{max} \end{cases} \tag{17}$$

The HIFU transducer modelled investigated in this paper was the same as in [12]. It is a spherically shaped bowl populated with $N = 256$ plane circular elements mounted onto its surface. The elements are each of $a = 3$ mm radius and of curvature $D = 18$ cm. The outer diameter of the HIFU transducer is 16 cm. The elements are pseudo-randomly spatially distributed on the surface of the array. The excitation frequency was assumed to be 1 MHz.

The rib topology was obtained from a human cadaver in STereoLithography format. Ribs 9-12 on the right side were truncated from the spine and closed surfaces were fitted over each rib using the Geomagic® software. CATIA v5 Advanced Meshing Tools were used to mesh the surfaces.

Within the context of this work, $q_{v_{max}}$ was chosen as 30% of the maximum value of the acoustic dose rate on the surface of the ribs resulting from all elements vibrating in phase and with 1 m s^{-1} amplitude velocity. This 30% value is not necessarily of clinical significance and was arbitrarily chosen to illustrate the technique. It is understood that in clinical applications, a damage threshold related to this dose rate quantity would have to be experimentally established

3. Results

The Helmholtz integral equation boundary element approach was implemented on a dedicated computer cluster and solved using Generalised Minimal Residual (GMRES) implementation. The approach used here differs from that described in [1] as follows.

- The speed of sound in the exterior domain was specified as a complex number in order to account for an amplitude absorption coefficient in the medium.
- Operations were vectorised to solve equation (6) for multiple right hand sides simultaneously, which enabled the coefficients of $[A]$ to be obtained more efficiently.
- The term in the surface Helmholtz integral equation containing the normal derivative of the acoustic pressure was accounted for in order to account for a locally reacting surface with a known specific admittance.

The routines employed to generate the boundary element matrices were obtained from the PAFEC (Program for Automatic Finite Element Calculations) VibroAcoustics software with permission from PACSYS Ltd.

A calculation on a simple scatterer was carried out for validation purposes. The results displayed here are for a 1 MHz unit amplitude plane wave scattered by a sphere with a complex surface impedance condition. The sphere was 10 mm in diameter and centred on the origin. These were compared against the analytical provided in [18] in the shadow zone between 5.5 mm and 25 mm (see figure 2).

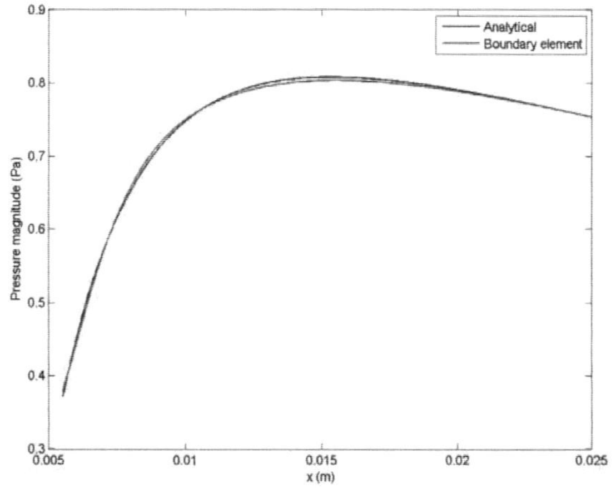

Figure 1. Helmholtz integral equation implementation of BE formulation on a spherical scatterer of 10 mm diameter. Incident field: unit amplitude 1 MHz plane wave travelling in positive x direction. Comparison along the x-axis against analytical solution.

The scatterer was meshed using isoparametric eight-noded quadratic patches ensuring at least three elements per wavelength for a wavespeed of 1500+4.405i m s^{-1}. The density of the medium was assumed to be 1000 kg m^{-3}. The properties of the scatterer were chosen to be representative of rib bone. The density of the scatterer was assumed to be 1912 kg m^{-3} and the speed of sound of longitudinal waves 4080 m s^{-1} [19]. The absorption coefficient of the scatterer was assumed to be 47.2 Np m^{-1} at 1 MHz [20]. Figure 1 shows agreement within 1.8% of the analytical solution on the x-axis in the shadow zone.

After completing this validation, the incident acoustic pressure field was replaced with that of the multi-element array described in [12] and a suitable mesh of the anatomical ribs described in Section 2 was generated. A simulation solving equation (6) for multiple right hand sides was launched so that the product $[H]^{-1}[\beta]$ could be obtained in four separate runs. The global origin was chosen to be at the focus of the array with propagation taking place along the positive z direction. This configuration is illustrated in plan view in figure 2.

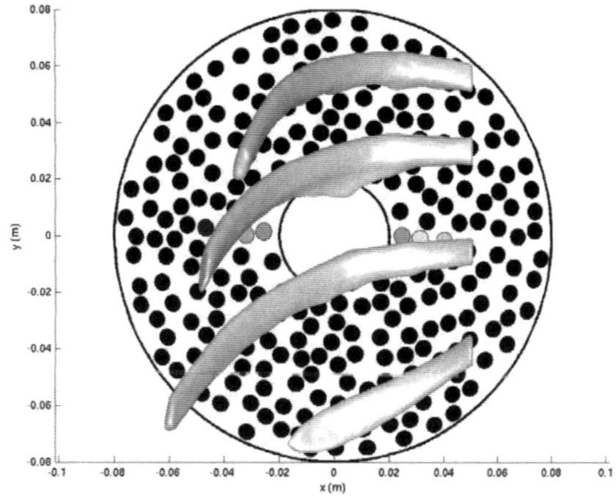

Figure 2. Position of ribs with respect to HIFU array looking in the negative z direction.

From a knowledge of $[H]^{-1}[\beta]$, the acoustic pressure on the surface of the ribs may be obtained using equation (9) for a given distribution of source velocities. In order to formulate the cost function and the constraints, the case of all elements vibrating in phase and with unit velocity amplitude was investigated. The incident acoustic pressure field at selected locations in the *y-z* plane in absence of ribs is shown in figure 3. The corresponding acoustic pressure field in the medium with the ribs present is shown in figure 4.

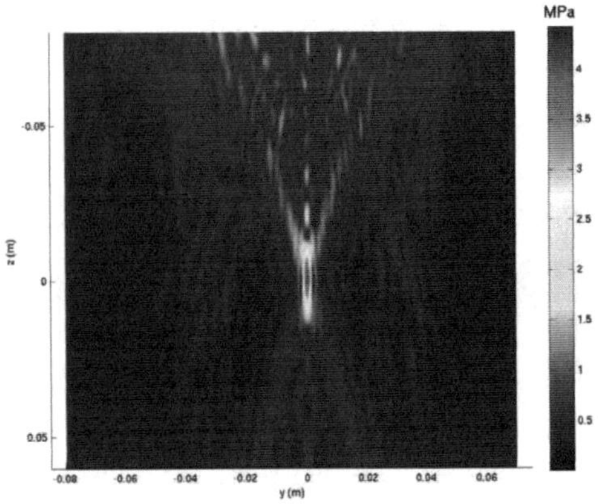

Figure 3. Incident acoustic pressure magnitude in *y-z* plane resulting from field of 1 MHz multi-element array. Uniform unit amplitude velocity and zero phase.

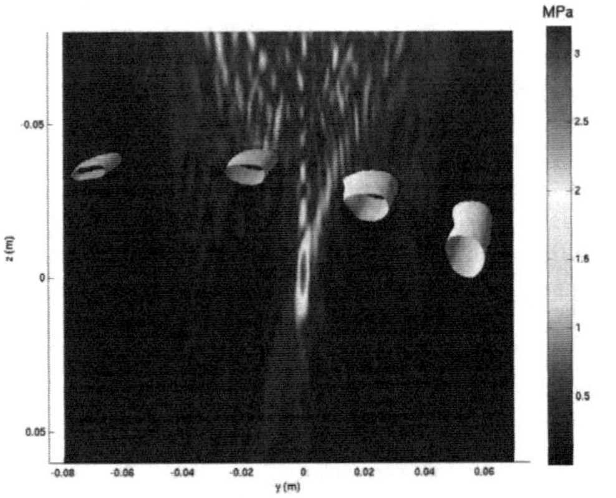

Figure 4. Acoustic pressure magnitude in *y-z* plane resulting from field of 1 MHz multi-element array. Uniform unit amplitude velocity and zero phase. Contour of ribs shown in bone colour.

The resulting acoustic dose rate on the surface of the ribs is shown in figure 5: this is shown in a plan view looking in the positive *z*-direction since the ribs are three-dimensional and their surface does not form a plane.

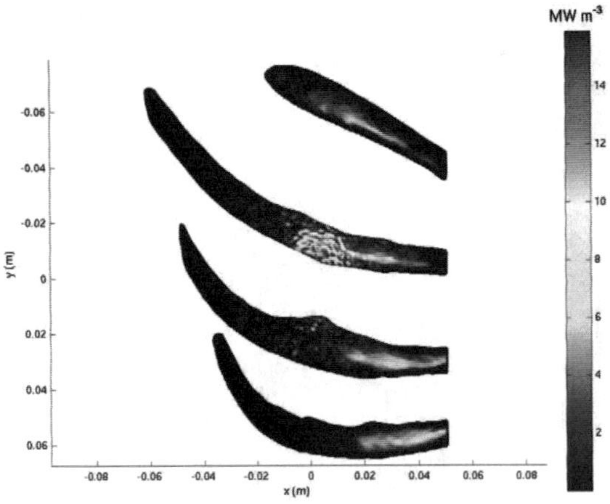

Figure 5. Acoustic dose rate on surface of ribs resulting from field of 1 MHz multi-element array. Spherical focusing (uniform unit amplitude velocity and zero phase).

Journal of Physics: Conference Series **498** (2014) 012004 doi:10.1088/1742-6596/498/1/012004

Figures 3 and 4 show that inserting the ribs between the array and the focus causes the acoustic pressure magnitude at the focus to drop from 4.4 MPa to 3.2 MPa. Figure 5 shows that the maximum acoustic dose rate value on the surface of the ribs is 16 MW m^{-3} (corresponding to an acoustic pressure of 1.6 MPa) for the spherical focusing case, with all elements of the array vibrating at an amplitude of 1 m s^{-1} and with uniform phase.

The NAG Numerical Library *e04us* routine was used to carry out the constrained minimisation [15]. It is designed to minimise an arbitrary smooth sum of squares function subject to constraints, which may include simple bounds on the variables, linear constraints and smooth nonlinear constraints. The procedure is described further in [12].

To fulfil the chosen requirement of a reduction to 30% in the rib surface acoustic dose rate compared to the spherical focusing case, the value $p_{surf_{max}}$ used to compute the acoustic dose rate in equation (17) was set to 0.88 MPa. U_{max} was selected to be 1 m s^{-1}. The vector of pressures in the exterior volume featured in the cost function (i.e. the 'desired' field pressure distribution) was generated from pressure field values in absence of ribs at 9261 equally spaced locations in a 21×21×21 cubic Cartesian grid such that -1.5 cm $\leq x \leq$ 1.5 cm, -1.5 cm $\leq y \leq$ 1.5 cm and -1.5 cm $\leq z \leq$ 1.5 cm, the focus of the array being at the global origin.

The surface mesh of the ribs contains 200921 nodes. It is both impractical and unnecessary to impose a constraint at all these locations on the surface of the scatterer. Too high a number of constraints may result in numerical instability and some locations on the surface of the ribs are highly unlikely to exceed $p_{surf_{max}}$, in particular those not directly facing the array. Hence, surface locations associated with pressure magnitudes below 40% of the maximum pressure correspond to the acoustic dose rate quantity in figure 5 (i.e. locations on the rib surface where the pressure was less than 0.64 MPa) were not included in the constraints. This resulted in only 3390 constraints associated with the pressure magnitude on the ribs being required, along with the 256 constraints for the magnitude of the element velocities. It will be verified after the optimisation that all nodes on the surface of the ribs are below the chosen pressure magnitude threshold.

With the above input conditions, a solution was obtained. The real and imaginary parts of the velocities were subsequently rescaled and the surface and field pressures calculated. Figures 6a and 6b respectively display the magnitudes and phases of the element velocities of the array resulting from the constrained minimisation. It was verified that the velocity magnitudes did not exceed 1 m s^{-1}.

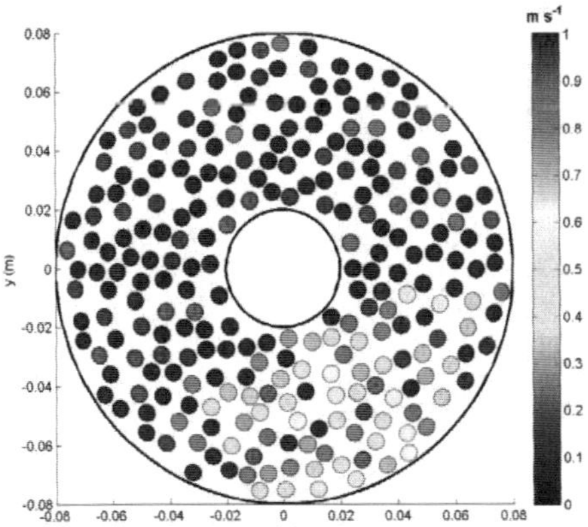

Figure 6a. Source velocity magnitudes resulting from constrained minimisation.

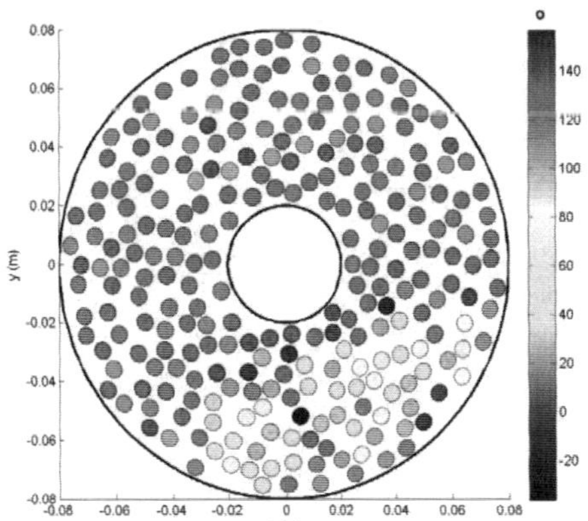

Figure 6b. Source velocity phases resulting from constrained minimisation.

Figure 7 shows the acoustic pressure magnitude in the *y-z* plane resulting from the source velocity distribution displayed in figures 6a and 6b.

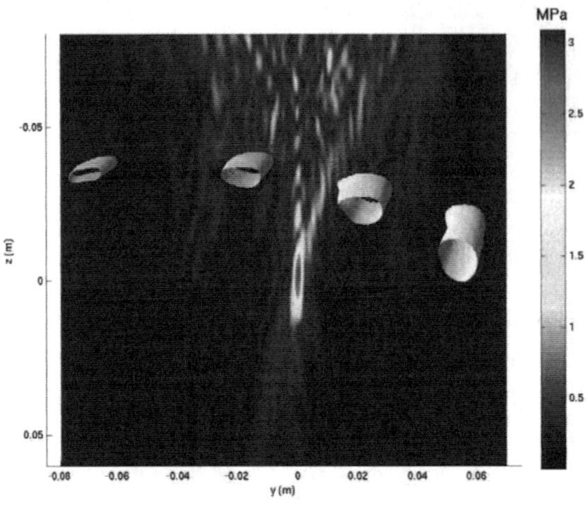

Figure 7. Acoustic pressure magnitude in *y-z* plane resulting from field of 1 MHz multi-element array. Source velocity distribution obtained from constrained minimisation.

Figure 7 shows that an overall reduction in the back-scattered acoustic pressure is achieved as a consequence of the constrained minimisation, when compared against the uniform amplitude and phase results displayed in figure 4. Although the constrained minimisation algorithm is unsuccessful at rendering a peak pressure of 4.4 MPa which was obtained in the absence of ribs (Figure 3), it is now 3.1 MPa, only 3% lower than in the case of spherical focusing in the presence of ribs (Figure 4).

Figure 8 shows the acoustic dose rate on the surface of the ribs, resulting from the source velocity distribution displayed in figures 6a and 6b.

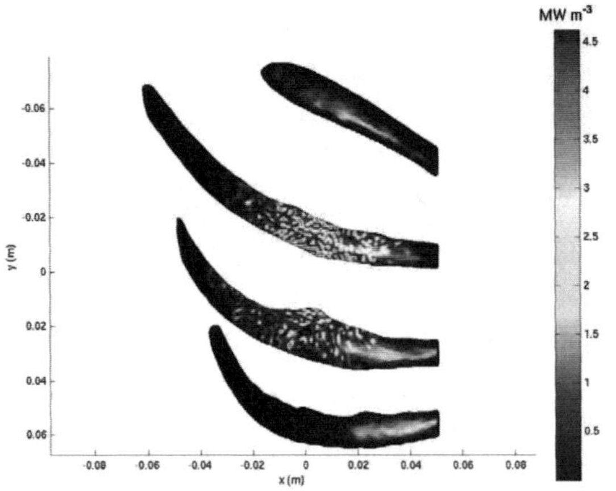

Figure 8. Acoustic dose rate on surface of ribs resulting from field of 1 MHz multi-element array. Source velocity distribution obtained from constrained minimisation.

The maximum acoustic dose rate on the bone side of the surface of the ribs is now 4.7 MW m^{-3}, corresponding to a maximum acoustic pressure of 0.88 MPa. A reduction in the maximum rib surface acoustic dose rate to 30% is therefore achieved with an ensued reduction in peak focal pressure compared with the spherical focusing case of only 3%.

4. Conclusions

An approach based on a GMRES implementation of the Helmholtz integral equation boundary element formulation was validated to model the scattering of the field of a multi-element random HIFU phased array by ribs of an arbitrary geometry under continuous wave excitation. The medium surrounding the scatterer featured a complex speed of sound thus allowing a phenomenological approach to modelling absorption effects. A surface acoustic impedance boundary condition was implemented to account for the fact that the ribs are not perfectly rigid and will also absorb acoustic energy. This enabled the calculation of an acoustic dose rate parameter on the surface of the ribs defined as the time-averaged relaxational absorption per unit volume [13].

Using the discretised form of the Helmholtz integral equation for locations in the exterior volume and on the surface of the scatterer, the inverse problem of determining the complex velocities of a multi-element array which produces an acoustic pressure field that best fits a required acoustic pressure distribution in a least-squares sense was formulated such that:

- The acoustic dose rate on the surface of the scatterer did not exceed a specified threshold, defined as 30% of the maximum acoustic dose rate resulting from spherical focusing.
- The amplitude of the velocity of each element on the array was bounded by maximum value specified as 1 m s^{-1}.

This approach was tested on a 1 MHz 256-element spherical array in presence of anatomical ribs. Employing a NAG library solver, a least-squares minimisation with nonlinear constraints was carried out to solve the above problem, where the gradients of the objective function and of the constraints with respect to the optimisation variables were provided. The solver returned a set of real and imaginary parts for the source velocities which satisfied both sets of constraints, hence reducing the back-scatterered pressure from the ribs and the acoustic dose rate on the surface of the scatterer compared with the spherical focusing case.

Further validation of the constrained minimisation methodology on other rib topologies is currently underway. The technique is also being benchmarked against binarised apodisation based on ray tracing and against the DORT method. This will form the topic of a future publication.

Acknowledgements

The authors would like to acknowledge the support of the National Measurement Systems Directorate of the UK Department of Business, Innovation and Skills and EPSRC EP/F025750/1. This work was supported in parts by the European Metrology Research Programme (Joint Research Project HLT03, which is jointly funded by the EMRP participating countries within EURAMET and the European Union) and by the Acoustics and Ionizing Radiation programme of the UK National Measurement Office. The authors are grateful to PACSYS Ltd. for supplying stand-alone executables of the Program for Automatic Finite Element Calculations boundary element routines as well as for useful discussions. The authors would equally like thank Minh Hoang for advice on inverse methods and help with NAG Numerical Library solvers. The authors would like to acknowledge the support of Tristan Clarke for IT-related matters. They would also like to express their gratitude to David Sinden, Ian Rivens, Simon Arridge and Timo Betcke for useful discussions

References

[1] Gélat P, ter Haar G and Saffari N 2011 Modelling of the acoustic field of a multi-element HIFU array scatterred by human ribs *Phys. Med. Biol.* **56** 5553.
[2] Crum L, Bailey M, Hwang J H, Khokhlova V and Sapozhnikov 2010 Therapeutic ultrasound: Recent trends and future perspectives *Physics Procedia* **3** 25
[3] Leslie T, Ritchie R, Illing R, ter Haar G, Phillips R, Middleton M, Wu F and Cranston D 2012 High-intensity focused ultrasound treatment of liver tumours: post-treatment MRI correlates

well with intra-operative estimates of treatment volume *Brit. J. Radiol.* **85** 1363–1370.

[4] Wu F, Zhi-Biao W, Wen-Zhi C, Hui Z, Jin B, Jian-Zhong Z, Ke-Quan L, Cheng-Bing J, Fang-Lin X, Hai-Bing S 2004 Extracorporeal high intensity focused ultrasound ablation in the treatment of patients with large hepatocellular carcinoma *Ann. Surg. Oncol.* **11** 1061

[5] Li J-L, X-Zh Liu, Zhang D, Gong X-F 2007 Influence of ribs on nonlinear sound field of therapeutic ultrasound. *Ultrasound Med. Biol.* **33** 1413

[6] Liu H-L, Chang Hsu, Chen W-S, Shih T-C, Hsiao J-K and Lin W-L. Feasibility of transrib focused ultrasound thermal ablation for liver tumours using a spherically curved 2D array: A numerical study 2007 *Med. Phys.* **34** 3436

[7] Quesson B, Merle M, Roujol S, de Senneville B D and Moonen C T. A method for MRI guidance of intercostal high intensity focused ultrasound ablation in the liver 2010 *Med. Phys.* **37** 2533

[8] Marquet F, Aubry J F, Pernot M, Fink M and M Tanter. Optimal transcostal high-intensity focused ultrasound with combined real-time 3D movement tracking and correction 2011 *Phys. Med. Biol.* **56** 7061

[9] Prada C, 2002 Detection and imaging in complex media with the DORT method *Top. Appl. Phys.* **84**, 107

[10] Cochard E, Aubry J-F, Tanter M and Prada C 2011 Adaptive projection method applied to three-dimensional ultrasonic focusing and steering through the ribs *J. Acoust. Soc. Am.* **130** 716

[11] Ballard J, Casper A J, Wan Y and Ebbini E S 2010 Adaptive Transthoracic Refocusing of Dual-Mode Ultrasound Arrays *IEEE T. Bio-med. Eng.* **57** 93

[12] Gélat P, ter Haar G and Saffari N 2012 The optimisation of acoustic fields for ablative therapies of tumours in the upper abdomen *Phys. Med. Biol.* **57** 8471.

[13] Duck F A 2009 Acoustic dose and acoustic dose rate *Ultrasound Med. Biol.* **35** 1679

[14] Nyborg W L 1981 Heat generation by ultrasound in a relaxing medium *J. Acoust. Soc. Am.* **70** 310

[15] http://www.nag.co.uk/numeric/numerical_libraries.asp

[16] Courant R and Hilbert D 1962 *Methods of Mathematical Physics* Vol. 2 (New York: Interscience; 1962), p258

[17] Colton D and Kress R 1983 Integral Equation Methods in Scattering Theory (Pure and Applied Mathematics) (New York: Wiley), p81 and p84

[18] Morse P M and Ingard K U 1968 *Theoretical Acoustics* (New York: McGraw-Hill), p422

[19] Wein W, Brunke S, Khamene A, Callstrom M R and Navab N 2008 Automatic CT-ultrasound registration for diagnostic imaging and image-guided intervention *Med. Image Anal.* **12** 577

[20] El-Brawany M A, Nassiri D K, Ter Haar G, Shaw A, Rivens I and Lozhken K 2009 Measurement of thermal and ultrasonic properties of some biological tissues *J. Med. Eng. Technol.* **33** 249.

Journal of Physics: Conference Series **498** (2014) 012005 doi:10.1088/1742-6596/498/1/012005

Multiple scattering in dispersions, for long wavelength thermoacoustic solutions

T A Hazlehurst[1], O G Harlen[1], M J Holmes[2] and M J W Povey[2]
[1]Department of Applied Mathematics, University of Leeds, Leeds, LS2 9JT, UK
[2] School of Food Science and Nutrition, University of Leeds, Leeds, LS2 9JT, UK
E-mail: mm07tah@leeds.ac.uk

Abstract. Thermoacoustic scattering is a principal scattering mechanism in the ultrasonic characterisation of water-continuous colloids. Thermal effects are particularly important in highly concentrated systems, where non-propagational thermal fields surrounding the disperse particles overlap. For low concentrations, the single sphere solution of Epstein and Carhart has become a popular tool for determining the particle size distribution. However, for small particle sizes it suffers from ill-conditioning that can make the solution numerically unstable. This problem has been resolved, by Harlen *et. al.* (2001, SIAM Journal on Applied Mathematics, **61** 1906–1931), who obtained an asymptotic solution for low concentrations that is valid when the particle diameter is small compared to the wavelength. In this paper we will use this asymptotic method to calculate the effects of multiple scattering that occur at higher concentrations. We use the addition translation theorem to calculate the effects of multiple scattering between a pair of spheres of different sizes and show how this affects the close-field scattering pattern.

1. Introduction

Ultrasonic techniques are becoming increasingly popular tools for characterising soft solids and colloids. In particular, ultrasound spectroscopy can be used to measure non-intrusively particle sizes within opaque materials in food, chemical and pharmaceutical manufacturing, where other techniques, such as, light scattering, are ineffective.

The first theory of ultrasound propagation in dispersions was developed by Lord Rayleigh [1]. This theory was then extended to fog and solid particles by Epstein and Carhart [2] and Allegra and Hawley [3], respectively. In thermoacoustic scattering the sound wave is scattered by the differences in thermal properties between the phases. The transmission problem reduces to the solution of a pair of coupled Helmholtz equations for the acoustic and thermal scattering fields. The acoustic field describing the scattered sound, with wavenumber k_c, while the thermal field represents the temperature fluctuations near the droplet boundaries and has a complex wavenumber k_T. This thermal wave decays rapidly with distance and the inverse of $|k_T|$ represents the distance heat can diffuse over the period of the radiation.

The ECAH (Epstein and Carhart and Allegra and Hawley) calculation does not account for multiple scattering within a suspension and so only provides accurate solutions for low concentrations. Lloyd and Berry [4] developed a method of including multiple scattering based on an infinite number of energy shells. However this method is only applied to the acoustic field and so does not consider the changes to the scattering from the overlap of the thermal fields between neighbouring droplets. Gaunaurd, Huang and Strifors [5] have formulated a method for calculating the scattering from a pair of spheres of different sizes at arbitrary angles to the incidence wave. They make use of the translation addition theorem to determine the secondary scattering of the scattered acoustic wave from a neigbouring sphere.

For the case of the scattering of a plane wave by a single spherical droplet, an analytical solution of the ECAH can be obtained as an infinite sum of spherical Bessel and Hankel functions. However, in the long acoustic wavelength limit $k_c a \ll 1$ (where a is the particle radius), typical of many colloids, the solution for the coefficients is ill-conditioned [6]. However by treating $k_c a$ as a small parameter, a regular perturbation solution can be found that converges rapidly on the solution [6–8].

In this paper we use this perturbation approach to address the multiple thermoacoustic scattering between a pair of spherical droplets. In section 4, we consider how the thermal fields interact when the spheres are within a distance of $1/|k_T|$ of one another. This is then used this to correct the far-field scatting pattern from which the attenuation is calculated. In section 5 we discuss the how the accuracy of the coefficients of the far-field scatting pattern depends upon the number of terms used in the near field expansions, particularly when the spheres are close.

2. Motivation

The motivation behind this work is the need to develop a theory for thermoacoustic scattering in colloids that is accurate for high concentrations. In figure (1) we compare the attenuation predicted by the ECAH theory with that measured in monodisperse silicon oil-in-water emulsions for concentrations ranging from 5 to 50 %. While the model is reasonably accurate at the lower concentrations. For concentrations above 20% the theory substantially over-predicts the attenuation. Furthermore, applying the correction of Lloyd and Berry does not correct this.

The explanation of this discrepancy can be found by examining the decay length of the thermal field around the oil droplets, which is given by [9] as

$$\delta_T = \sqrt{\frac{2\tau}{\rho C_p \omega}}, \tag{1}$$

where τ, ρ, C_p and ω are thermal conductivity, density, specific heat capacity and angular frequency respectively. At low concentrations the droplets are sufficiently far enough apart that these fields don't interact. However if

$$\delta_T \geq \frac{d}{2}, \tag{2}$$

where d is the distance between the centre of neighbouring droplets, then the thermal field of two droplets will overlap. From this we can define the critical concentration ϕ_C at which the thermal fields between neighbouring particles will overlap. This critical volume fraction is given by [10]

$$\phi_C = \left(\frac{1}{1+\frac{\delta_T}{a}}\right)^3, \tag{3}$$

which for the case of the experimental in figure (1) is approximately 18%.

3. Low Frequency Thermoacoustic scattering

In thermoacoustic scattering the scattered field is written as the sum of an acoustic and a thermal wave so that the velocity potential is given by $e^{ik_c z} + \varphi + \psi$ in the continuous phase and $\varphi' + \psi'$ in the suspended phase, where $e^{ik_c z}$ is the incident plane wave, φ and φ' are the

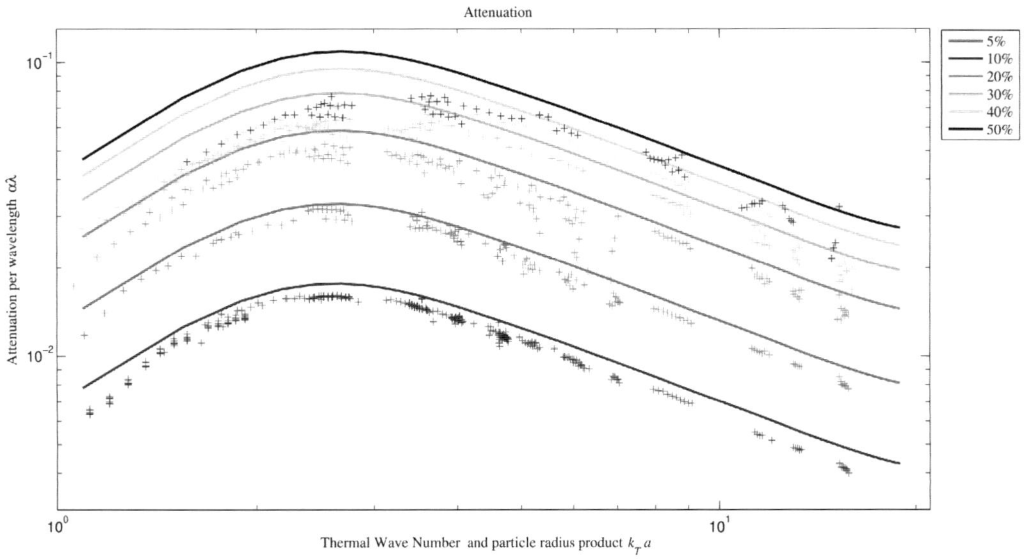

Figure 1. Attenuation spectrum based on the ECAH approach for a silicone oil-in-water emulsion of particle radius $0.23\mu m$ for different concentrations. It can be seen that when the concentration rises above the critical volume fraction ϕ_C given in (3) (in this case $\phi_C = 18\%$) the theory significantly overpredicts the attenuation. Experimental data from Herrmann [11].

scattered acoustic fields in the two phases and ψ and ψ' the thermal fields. Each of the four fields satisfies a Helmholtz equations [2],

$$\left(\nabla^2 + k_c^2\right)\varphi = 0, \quad \left(\nabla^2 + k_T^2\right)\psi = 0 \quad \text{in the continuous phase,} \tag{4}$$

and

$$\left(\nabla^2 + k_c^{'2}\right)\varphi' = 0, \quad \left(\nabla^2 + k_T^{'2}\right)\psi' = 0 \quad \text{in the droplet phase,} \tag{5}$$

where the acoustic and thermal wave numbers are given by $k_c = \frac{\omega}{c}(1 + i\alpha)$ and $k_T = (1+i)\sqrt{\frac{\omega}{2\sigma}}$, respectfully. Here c, α and σ are the wave speed, attenuation coefficient and the thermometric conductivity.

The thermal and acoustic waves are coupled through the boundary conditions at the phase boundaries, where we have continuity of:

Normal velocity

$$\frac{\partial}{\partial n}\left(e^{ik_c z} + \varphi + \psi\right) = \frac{\partial}{\partial n}\left(\varphi' + \psi'\right), \tag{6}$$

Pressure

$$e^{ik_c z} + \varphi + \psi = \hat{\rho}(\varphi' + \psi'), \tag{7}$$

Temperature

$$\Gamma_c\left(e^{ik_c z} + \varphi\right) + \Gamma_t\psi = \Gamma_c'\varphi' + \Gamma_t'\psi', \tag{8}$$

Heat flux

$$\Gamma_c \frac{\partial}{\partial n}\left(e^{ik_c z} + \psi\right) + \Gamma_t \frac{\partial}{\partial n}\psi = \hat{\tau}\left(\Gamma'_c \frac{\partial}{\partial n}\varphi' + \Gamma'_t \frac{\partial}{\partial n}\psi'\right), \tag{9}$$

where $\Gamma_c, \Gamma'_c, \Gamma_T$ and Γ'_T are as described in [8].

In most applications the radius of colloidal particles is much smaller than the length of the acoustic waves, so that $|k_c a| \ll 1$. For this low frequency limit we can follow the approach of Harlen *et al.* [8], and find a pertubation solution for the limit when the acoustic wavelength is large compared to droplet radius, $k_c a \ll 1$ but where the width of the thermal δ_T is comparable with particle radius so that$|k_T a| \sim 1$.

The external Helmholtz equation (4) is first transformed by defining,

$$\tilde{\varphi} = e^{-ik_c(r-a)}\varphi. \tag{10}$$

where r is radial distance from the origin, to a give a regular problem at infinity, allowing us to seek regular perturbation solution as power series in $ik_c a$,

$$(\tilde{\varphi}, \varphi', \psi, \psi') = \sum_{n=0}^{\infty} (ik_c a)^n (\tilde{\varphi}_n, \varphi', \psi, \psi') \tag{11}$$

This series converges provided $|k_c a| < \ln(2)$ [13] and converges rapidly, giving accurate answers with just a few terms, when $|k_c a| << 1$ [8]. Substituting the expansion into the Helmholtz equations (4) gives the following system, for $n \geq 0$,

$$\nabla^2 \tilde{\varphi}_n = -\frac{2}{ar}\frac{\partial}{\partial r}(r\tilde{\varphi}_{n-1}), \quad \text{for the continuous phase} \tag{12}$$

and

$$\nabla^2 \varphi'_n = \left(\frac{k'_c}{k_c}\right)^2 \varphi'_{n-2}, \quad \text{for the dispersed phase.} \tag{13}$$

The thermal fields remain given by Helmholtz equations with

$$\left(\nabla^2 + k_T^2\right)\psi_n = 0, \quad \text{in the continuous phase} \tag{14}$$

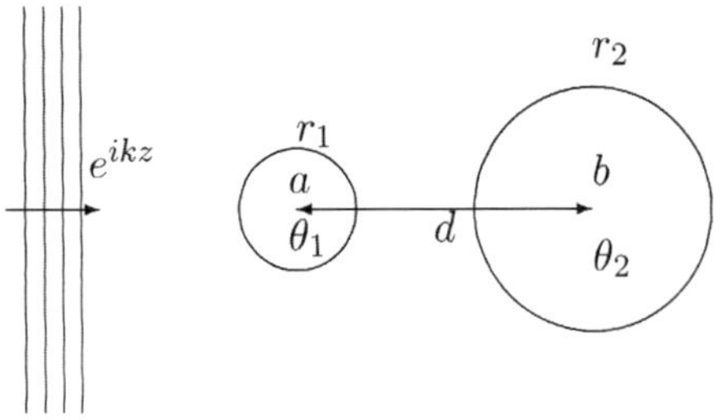

Figure 2. Set up for the two particle problem with a plane wave travelling parallel to the placement of two different sized spheres of radii a and b respectively separated by a distance d. We have a plane wave described by $e^{ik_c z}$ where z is the coordinate along the axis of the particle centres and r_1, θ_1 and r_2, θ_2 denote the polar coordinates with the origin based on centres of the spheres 1 and 2 respectively.

and

$$\left(\nabla^2 + k_T'^2\right)\psi_n' = 0, \quad \text{in the dispersed phase} \tag{15}$$

The boundary conditions at the phase boundaries can be grouped into pairs. The pressure and normal velocity boundaries give one pair relating $\tilde{\varphi}_n, \varphi_n', \psi_n$ and ψ_n', where the thermal conditions gives relations between $\tilde{\varphi}_{n-2}, \varphi_{n-2}', \psi_n$ and ψ_n', where $\psi_0 = \psi_1 = \psi_0' = \psi_1' = 0$. From this solutions can be calculated sequentially in the order, $\{\tilde{\varphi}_0, \varphi_0'\}$,$\{\tilde{\varphi}_1, \varphi_1'\}$, $\{\psi_2, \psi_2'\}$, $\{\tilde{\varphi}_2, \varphi_2'\}$, $\{\psi_3, \psi_3'\}, \ldots$. At order $n = 0$ the solution is simply $\{\tilde{\varphi}_0 = 0, \varphi_0' = 1/\hat{\rho}\}$.

4. Low frequency thermoacoustic scattering by two spherical droplets

We now consider the thermoacoustic scattering solutions for two spheres, which for simplicity we assume are arranged parallel to the direction of the incident wave, separated by a distance d, as shown in figure (2). We use two polar coordinate systems, denoted by (r_1, θ_1) and (r_2, θ_2) centred on particles 1 and 2 respectively.

Following the method of Gaunaurd [5], we express the thermal wave in the continuous phase as the sum of two spherical wave expansions centred on the particle centres. Since we have chosen the particles to lie along the axis of the incident wave the solution is axisymmetric so that we only need to retain the $p = 0$ terms (the more general case is given in reference [5]), so that

$$\psi_n = \sum_{q=0}^{\infty}\left[A_{nq}h_q^{(1)}(k_T r_1)P_q(\cos\theta_1) + B_{nq}h_q^{(1)}(k_T r_2)P_q(\cos\theta_2)\right] \tag{16}$$

As we also need to consider the boundaries around both particles, the solution needs to be expressed solely in terms of r_i and θ_i, where $i = 1, 2$ depending on whether we are considering solutions about the first sphere, (1), say, or the second sphere, (2). To do this we use the translation theorem to express the solution in terms of coordinates centred about particle A,

$$\psi_n^{(1)} = \sum_{q=0}^{\infty}\left[C_{nq}h_q^{(1)}(k_T r_1) + D_{nq}\sum_{m=0}^{\infty} Q_{T(qm)}^{(2)}(d) j_q(k_T r_1)\right] P_q(\cos\theta_1), \tag{17}$$

where

$$Q_{T(qm)}^{(2)}(d) = i^{q-m}(2q+1)\sum_{\sigma=|q-m|}^{q+m} i^{\sigma} b_{\sigma}^{(qm)} h_{\sigma}^{(1)}(k_T d), \tag{18}$$

and

$$b_{\sigma}^{(qm)} = (2\sigma+1)\begin{pmatrix} q & m & \sigma \\ 0 & 0 & 0 \end{pmatrix}^2, \tag{19}$$

where the Wigner $3-j$ symbols are defined by [12]. Similarly, about the second particle (2),

$$\psi_n^{(2)} = \sum_{q=0}^{\infty}\left[D_{nq}h_q^{(1)}(k_T r_2) + C_{nq}\sum_{m=0}^{\infty}(-1)^{q+m} Q_{T(qn)}^{(2)}(d) j_q(k_T r_2)\right] P_q(\cos\theta_2). \tag{20}$$

The thermal fields inside particle (i) are given by

$$\psi_n'^{(i)} = \sum_{q=0}^{\infty} c_{nq}^{(i)} j_q(k_T r_i) P_q(\cos\theta_i). \tag{21}$$

The solution for the external acoustic field can be written as

$$\tilde{\varphi}_n = \sum_{q=0}^{\infty} A_{nq}\left(\frac{a}{r_1}\right)^{q+1} P_q(\cos\theta_1) + \sum_{q=0}^{\infty} B_{nq}\left(\frac{b}{r_2}\right)^{q+1} P_q(\cos\theta_2) + I_n, \tag{22}$$

where I_n is a particular solution of the inhomogenous equation and for $n = 0$ and $n = 1$, $I_n = 0$. Again we can use the translation theorem to express this in terms of the coordinate system around particle (1) as

$$\tilde{\varphi}_n^{(1)} = \sum_{q=0}^{\infty} \left[A_{nq} \left(\frac{a}{r_1}\right)^{q+1} + B_{nq} \sum_{m=0}^{\infty} Q_{c(qm)}^{(2)} \left(\frac{r_1}{a}\right)^{q+1} \right] P_q(\cos\theta_1) + I_n, \tag{23}$$

and internally in sphere (1)

$$\varphi_n^{(1)} = \sum_{m=0}^{\infty} a_{nq}^{(1)} \left(\frac{r_1}{a}\right)^q P_q(\cos\theta_1) + I_n^{(1)}, \tag{24}$$

where $I_n^{(1)}$ is a particular solution of the inhomogeneous equation, where $I_0^{(1)} = I_1^{(1)} = 0$, $I_2^{(1)} = k_c'^2/k_c^2 a^2 \hat{\rho}$. Together with the equivalent equations centred on particle (2). Applying the boundary conditions at the particle interfaces gives a system of linear equations for the unknowns A_{nq} and B_{nq} or C_{nq} and D_{nq}.

5. Results

Solving these systems allows us to us to study the close field interactions between the particles, as seen in figures 3 through 6. In figure 3 we show thermal field for the case of a 200nm and a 1 μm particle. The black dashed lines show the extent of the thermal field δ_T around each particle. In this case the thermal field of the smaller particle is engulfed by the field of the larger particle. Consequently the scattering is dominated by the larger particle.

In figure 4 to 6 we show how the thermal field around two particles of equal size (1 μm) varies as the particle separation changes. In figure 4 the proximity of the particles to each other means that there is heavy overlapping of the thermal fields. As a consequence the total thermal field appears equivalent to that of a single larger particle. This reduces the acoustic attenuation from that of two isolated particles.

The two particles are separated a little further apart in figure 5, with just a small overlap of the thermal fields. However, there are still significant particle-particle interactions mediated through the thermal field. Finally, we can see in figure 6 the particles are much further apart. In this case the thermal field is approximately equal to that of the superposition of the thermal fields from two isolated particles so that the two particles are not coupled thermally.

One issue with solving the linear systems given by applying the boundary conditions to the expansions given in equations (17) and (23) is the number of terms in q required to obtain an accurate solution. Figures 7 and 8 show the errors in the terms A_{20} and A_{22} of the acoustic field respectfully, which are the terms that dominate the far field scattering pattern. When the particles are well separated only a small number of terms are required to give an accurate answer. In these cases the coupling is through the acoustic field. As expected, it can be seen that when the particles are closer together that a larger number of terms are required to give an accurate answer. However, even in the limit $d = 2a$ when particles are in contact with one another the series converge and can be computed with a moderate number of terms.

6. Conclusion

We have demonstrated how the perturbation solution for solving thermoacoustic scattering in the long wave limit can be applied to the scattering by two closely separated spherical particles. This allows us to examine the extent of overlapping of the thermal fields and hence estimate the degree to which the overlapping of thermal fields reduces the thermoacoustic attenuation in a concentrated colloid. The next step is to generalise these results to multiparticles at a

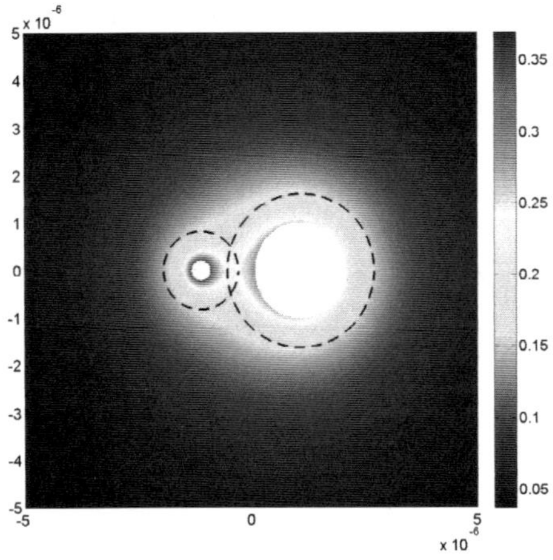

Figure 3. The thermal field of two particles, $a = 2{\cdot}10^{-7}m$, $b = 1{\cdot}10^{-6}m$ and $d = 2b$. The black circles show the thermal wave decay described in (1).

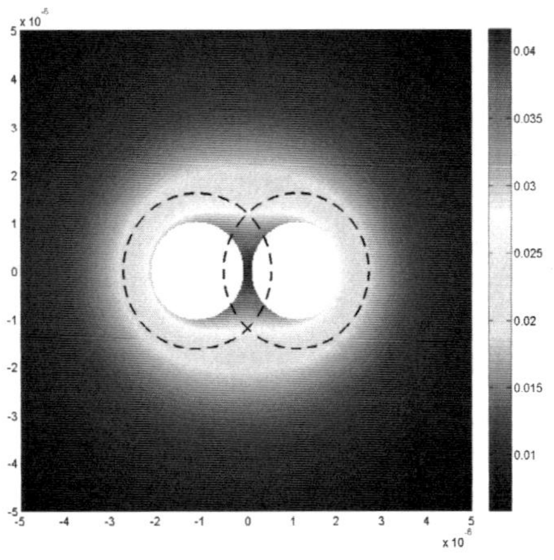

Figure 4. The thermal field of two particles, $a = 1 \cdot 10^{-6}m$, $b = 1 \cdot 10^{-6}m$ and $d = 2.2a$. The black circles show the thermal wave decay described in (1).

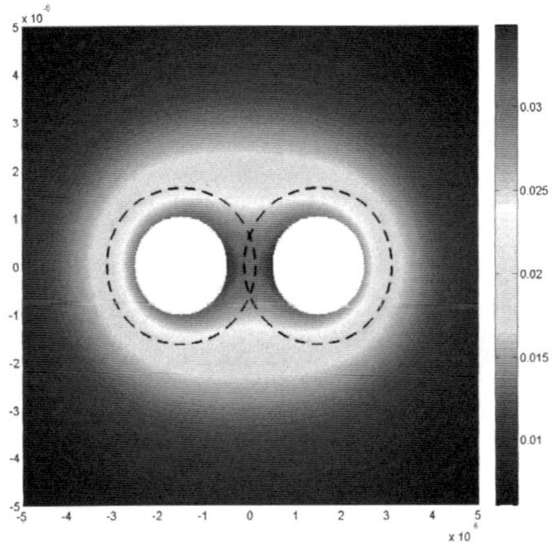

Figure 5. The thermal field of two particles, $a = 1{\cdot}10^{-6}m$, $b = 1{\cdot}10^{-6}m$ and $d = 3a$. The black circles show the thermal wave decay described in (1).

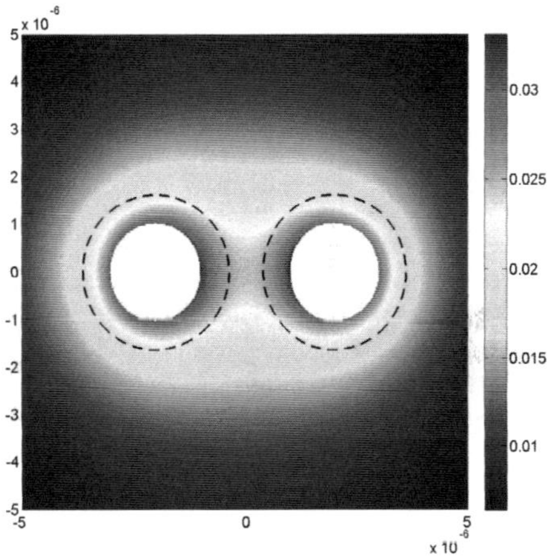

Figure 6. The thermal field of two particles, $a = 1{\cdot}10^{-6}m$, $b = 1{\cdot}10^{-6}m$ and $d = 6a$. The black circles show the thermal wave decay described in (1).

general angle of incidence in order to quantify these effects. This will provide a method for more accurately describing acoustic propagation in more higher concentration colloids than is given by current theories.

7. Acknowledgement
We are grateful to the UK Engineering and Physical Sciences Research Council (EPSRC) and Procter and Gamble for funding for this work through a CASE studentship for TAH.

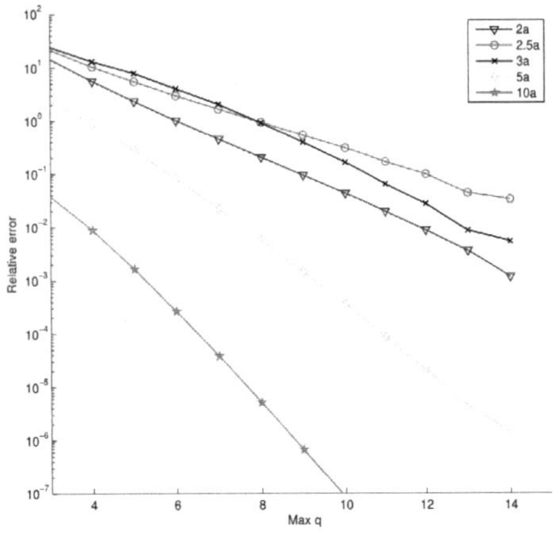

Figure 7. The error of A_{20} as a function for different distanced between the particles.

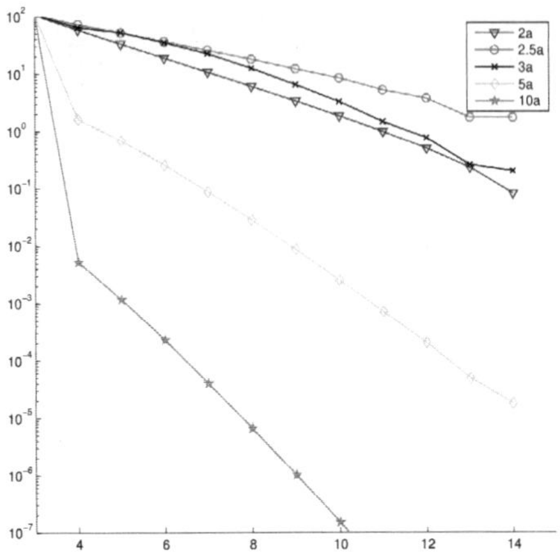

Figure 8. The error of A_{22} as a function for different distanced between the particles.

References

[1] Lord Rayleigh 1897 On the incidence of aerial and electric waves upon small obstacles in the form of ellipsoids or elliptic cylinders and on the passage of electric waves through a circular aperture in a conducting screen *Phil. Mag.* XLIV **266** 28-52

[2] Epstein P S and Carhart R R 1953 The absorption of sound in suspensions and emulsions I. water fog in air *J. Acoust. Soc. Am.* **25** 553-65

[3] Allegra J R and Hawley S A 1972 Attenuation of sound in suspensions and emulsions: theory and experiments *J. Acoust. Soc. Am.* **51** 1545-64

[4] Lloyd P and Berry M V 1967 Wave propagation through an assembly of spheres 4. relations between different multiple scattering theories *Proc. Phys. Soc.* **91** 678–688

[5] Gaunaurd G C, Huang H and Strifors H C 1995 Acoustic scattering by a pair of spheres *J. Acoust. Soc. Am.* **98** 495–507

[6] Harlen O G, Holmes M J, Povey M J W, Qiu Y and Sleeman B D 2001 A low frequency potential scattering description of acoustic propagation in dispersions *SIAM J. Appl. Math.* **61** 1906–1931

[7] Harlen O G, Holmes M J, Povey M J W and Sleeman B D 2003 Acoustic propagation in dispersions and the geometric theory of diffraction *SIAM J. Appl. Math.* **63** 834–849

[8] Harlen O G, Holmes M J, Pinfield V J, Povey M J W and Sleeman B D 2010 A perturbation solution for long wavelength thermoacoustic propagation in dispersions *J. Comput. Appl. Math.* **234** 1996–2002

[9] Pinfield V J and Povey M J W 1997 Thermal scattering must be accounted for in the determination of adiabatic compressibility *J. Phys. Chem. B* **101** 1110–1112

[10] Pinfield V J 1996 *Studies of Creaming, Occulation and Crystallisation in Emulsions: Computer Modelling and Analysis of Ultrasound Propagation* (PhD thesis, Procter Department of Food Science, University of Leeds)

[11] Hemar Y, Herrmann N, Lemarechal P, Hocquart R and Lequeux F 1997 Effective medium model for ultrasonic attenuation due to the thermo-elastic effect in concentrated emulsions *J. Phys-Paris II*, **7** 637–647

[12] Edmonds A 1957 *Angular Momentum in Quantum Mechanics* (Princeton U.P., London)

[13] Kleinmann R E 1965 The Dirichlet problem for the Helmholtz equation *Arch. Ration. Mech. Anal.* **18** 205–229

Journal of Physics: Conference Series **498** (2014) 012006 doi:10.1088/1742-6596/498/1/012006

Experimental measurements of the coherent field resulting from the interaction of an ultrasonic shock wave with a multiple scattering medium

Nicolas Viard, Bruno Gianmarinaro, Arnaud Derode, Christophe Barrière

Institut Langevin, 1 rue Jussieu, 75005 Paris, France

E-mail: nicolas.viard@espci.fr

Abstract. Whereas multiple scattering and shock wave formation are known to be antagonistic phenomena, this work concentrates on the interaction of an ultrasonic shock wave with a random multiple scattering medium. The shock wave is generated by long distance propagation of a short pulse (4 periods at a 3.5 MHz central frequency) in water before it encounters the scattering medium (a slab-shaped random set of parallel metallic rods). Transmitted waves are recorded over hundreds of positions along the lateral dimension of the slab to estimate the ensemble-averaged transmitted field $\langle\phi(t)\rangle$, also known as the coherent wave. Experiments are repeated for different thicknesses L of the slab and different emission amplitudes. The elastic mean free path l_e (i.e the typical distance for the decreasing of the coherent intensity $|\langle\phi(t)\rangle|^2$ due to scattering) is determined as well as the harmonic rate of the averaged transmitted wave. Experimental results are discussed and compared to the linear case.

1. Introduction

Acoustic wave propagation is intrinsically nonlinear and classical acoustic nonlinearities act as a cumulative effect [1]. For a plane wave propagating in a homogeneous non dissipative fluid, one usually defines the shock formation distance $L_s = (k\beta M)^{-1}$ as the distance the wave has to travel before a discontinuity appears in its pressure waveform. It is inversely proportional to the wavenumber $k = 2\pi/\lambda$, to the fundamental nonlinear parameter β, and to the Mach number $M = v_0/c_0$, where v_0 is the particle velocity and c_0 is the sound speed. Linear approximation is usually made if the travelled distance is much shorter than L_s. Otherwise, if the travelled distance is approximately equal or superior to L_s, the propagation regime is nonlinear. Multiple scattering of wave is a general phenomenon. It can concerns electrons, optical waves, acoustic waves as well as seismic waves for example. For the acoustic waves, it comes out that multiple scattering could be beneficial on as well as antagonistic towards the classical acoustic nonlinearities. Indeed, multiple scattering lengthens travelled paths in the medium which is beneficial on the cumulative nonlinearities, while it also spreads the energy in the medium, yielding to lower values of v_0 corresponding to higher values of L_s.

In this work, we investigate the interaction of an ultrasonic shock wave with a multiple scattering medium. Experimental measurements of the coherent wave field, i.e., the ensemble-averaged field, are compared for different emission amplitudes corresponding to linear and nonlinear regimes. The ultrasonic shock wave is formed in water before it reaches the

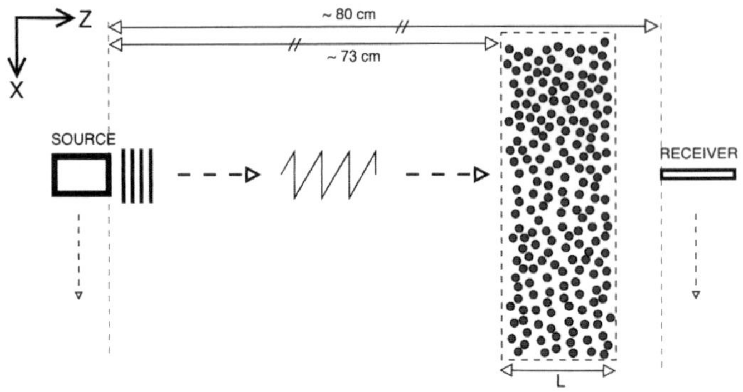

Figure 1. Experimental set-up. A short ultrasonic pulse around 3.5 MHz is emitted by a planar circular transducer (⌀ 38 mm). It propagates into water before it reaches the multiple scattering slab ($10 < L < 40$ mm). Transmitted scattered waves are recorded by an hydrophone (⌀ 200 μm).

heterogeneous medium. The latter consists of parallel steel rods randomly placed in water. Transmitted scattered waves are recorded and averaged to form the coherent wave. In the linear approach, the coherent wave has an effective wavenumber k_{eff} whose imaginary part accounts for losses due to scattering. As the wave propagates into the medium, the intensity of its coherent part decreases exponentially with a characteristic length $l_e = 1/(2\,\mathrm{Im}(k_{eff}))$ called the elastic mean free path. Harmonic components of the coherent wave are compared for different sample thicknesses. The elastic mean free path l_e is estimated as a function of frequency. Because of nonlinear propagation, the increase of the emission amplitude extends the frequency band and therefore allows to estimate the elastic mean free path at higher frequencies. At a given frequency, we observe that, estimates at high emission amplitudes do not change compare to esimates at a low amplitude, which implies that multiple scattering in this kind of medium is purely a linear phenomenon.

2. Experimental procedure

Experimental setup is depicted in figure 1. A 4-period tone burst at a 3.5 MHz central frequency with a gaussian amplitude modulation is generated by a Tektronix arbitrary waveform generator (AFG 3101) and amplified by an AR amplifier (75A250A, 75W, 10kHz-250Mhz). It is emitted by a planar circular transducer ($2r = 38$ mm, nominal central frequency $f_0 = 3.5$ MHz, $\Delta f/f_0 = 100\%$ at -6dB) in water, where the corresponding wavelength is 0.43 mm. The wave propagates over 73 cm in water ($c_0 = 1500$ m/s, $\beta = 3.5$) before it reaches the multiple scattering medium. It is a random set of 0.8 mm diameter steel rods with density 12 rods/cm^2. A set of slabs with different thicknesses allows us to vary the total thickness of the multiple scattering medium from 10 to 40 mm. As a reminder, the density of steel ρ_{steel} is 7800 kg/m^3, the longitudinal wavespeed c_L is 5.7 mm/μs and the transversal wavespeed c_T is 3 mm/μs. Using an HGL-200 hydrophone from ONDA, one records the scattered wave that are transmitted through the sample at a constant distance from the source, i.e., 80 cm. Such a distance ensures an optimized configuration in terms of spectral components and diffracted waveform. Indeed, at a shorter distance from the surface of the transducer, due to interferences between the direct and the edge waves from the transducer, the amplitude oscillates between zero and a maximal value along the axis z of the transducer. At a greater distance, because of diffraction, the ultrasonic beam diverges. It is commonly refered to the Rayleigh distance $d_R = \pi r^2/\lambda$, the distance from which the acoustic beam is no longer collimated. Here, $d_R > 2$ m. At last, it is important to compare the shock formation distance L_s to some characteristic length for the intrinsic attenuation, whose effect limits the creation of higher order harmonics. From [2], we obtain the characteristic length for the intrinsic attenuation in pure water, $L_a = 3.4$ m at 3.5 MHz. The shock formation distance L_s is calculated for the lowest and the highest emission amplitudes, respectively $v_0 = 0.002$ m/s and $v_0 = 0.05$ m/s, at $f = 3.5$ MHz. L_s is found equal to 14.6 m and 0.6 m. Therefore, the regime of propagation can be considered as highly nonlinear for the highest emission amplitude.

In order to estimate the coherent wave, the recording of the transmitted wave is repeated over a few hundreds positions along the medium by translating both the transducer and the hydrophone. Rigourosly the coherent wave -i.e., the ensemble averaged transmitted field- is given by the following equation :

$$\langle \phi(t) \rangle = \frac{1}{N} \sum_{i=1}^{N} \phi_i(t), \qquad N \to \infty \tag{1}$$

where i denotes the number of positions -or realisations of the disorder. Nevertheless, in our experimental configuration, ensemble-averaged of the transmitted field cannot be achieved because of the finite size of our sample. Therefore, the number of positions N is determined, both by the spatial step p and the length of the scan over the lateral dimension x. Spatial step p was determined so that two adjacent positions are uncorrelated, whereas the length of the scan was chosen in order not to be perturbed by the edges of the sample. With a spatial step $p = 0.5$ mm and a lateral scan over 20 cm, we obtain $N = 401$. When one scan is finished, the electrical power transmitted to the transducer is increased by changing the voltage of the generator, and a new scan begins.

Transmitted signals in water (without the scattering slab) are shown in figure 2. In the temporal domain, the waveform stiffens when the emission power is increased which is typical of the classical nonlinearities [3, 4, 5]. At the lowest emission power (1.6 mW), the transmitted signal only shows frequency components around f_0 which indicates that even if the travelled distance (i.e., $\sim$ 80 cm) is large, the regime of propagation stays linear. In comparison, at the highest emission power (1.1 W), frequency components spread at least until the 7th harmonic (i.e., 24.5 MHz).

For the determination of the frequency-resolved elastic mean free path, $l_e(f)$, a procedure from [6] is followed. The transmission coefficient for the energy of the coherent wave, T_c, is calculated for different thicknesses L of the slab on narrow frequency bands (width $\delta f = 0.2$ MHz). Then, the frequency-resolved elastic mean free path $l_e(f)$ is deduced from a linear fit of $\log(T_c(L, f))$ as a function of L, for each value of f.

3. Results and discussion

Estimates of the coherent wave at the lowest and the highest emission amplitudes are presented in figure 3 for two different thicknesses L. For the finest sample ($L = 10$ mm), the duration of the estimate of the coherent wave is really alike the duration of the measured signals in water, whereas for the sample with thickness $L = 30$ mm, the duration of the estimate of the coherent wave is much greater than the duration of the measured signals in water. This is due to a resonance in the scattering cross section of the steel rod occuring at the 2.75 MHz frequency [6]. At a given thickness, the two estimates of the coherent wave corresponding to the lowest and to the highest emission powers look really alike, excepted that the curve corresponding to the highest power is stiffened compare to the one corresponding to the lowest emission power.

Figure 4 shows the harmonic amplitudes for the first -fondamental-, the second, the third and the fourth harmonics. For every thickness L of the sample, the harmonic amplitude is plotted as a function of the emission amplitude (here, the square root of the electrical power transmitted

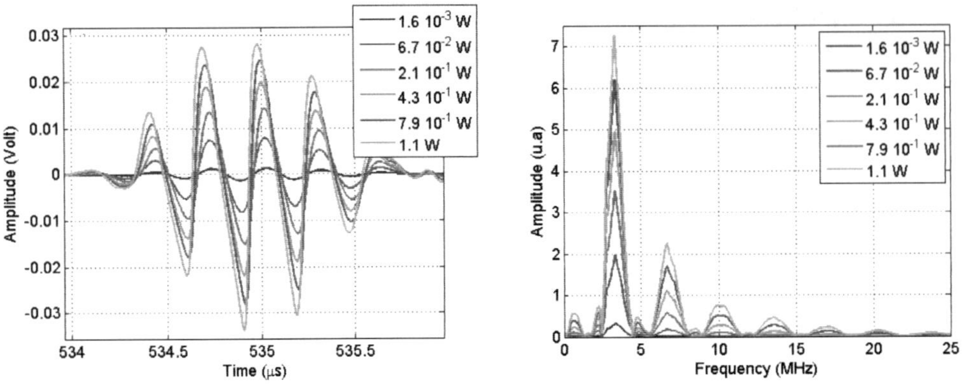

Figure 2. Measured signals after propagation in water over 80 cm for different emission powers. At the lowest emission power (i.e., 1.6 mW), the sprectrum has only components around 3.5 MHz, while, at higher power, other components successively appear around higher order harmonics: 7, 10.5, 14, 17.5 MHz...

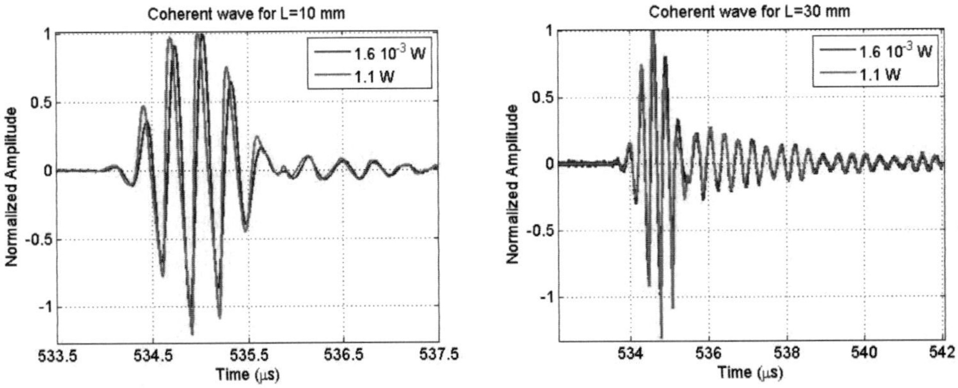

Figure 3. Estimates of the coherent wave ($N = 401$ realisations) for the lowest and the highest emission powers. The thickness of the sample is equal to 10 mm (left) and 30 mm (right).

to the transducer). Each curve is normalized by its value at the highest emission amplitude so that it can be compared to the others. It is remarkable to see that, at a given frequency, whatever is the thickness of the multiple scattering slab, all the curves corresponding to the different thicknesses and to the transmission in water ($L = 0$ mm) overlap merely perfectly. This indicates that in our experimental configuration, the multiple scattering does not change the behavior of one given harmonic or does not act nonlinearly.

As shown in figure 2, the order of the highest harmonic that is measurable depends on the emission power. For example, at a 0.21 W emission power, the spectrum components of the transmitted wave in water are limited around the first and the second harmonics, while at a 0.79 W emission power, spectrum components extend to the third harmonic. Therefore, depending on the emission amplitude, the elastic mean free path l_e is determined in a different frequency range.

In figure 5, the estimates of the elastic mean free path versus frequency are represented for different emission powers. Over the frequency interval 2.4-4.4 MHz the elastic mean free path is estimated with four sets of data corresponding to the following emission powers : 1.6 mW, 0.21, 0.79 and 1.1 W. The four estimates are in very good agreement with each other. They

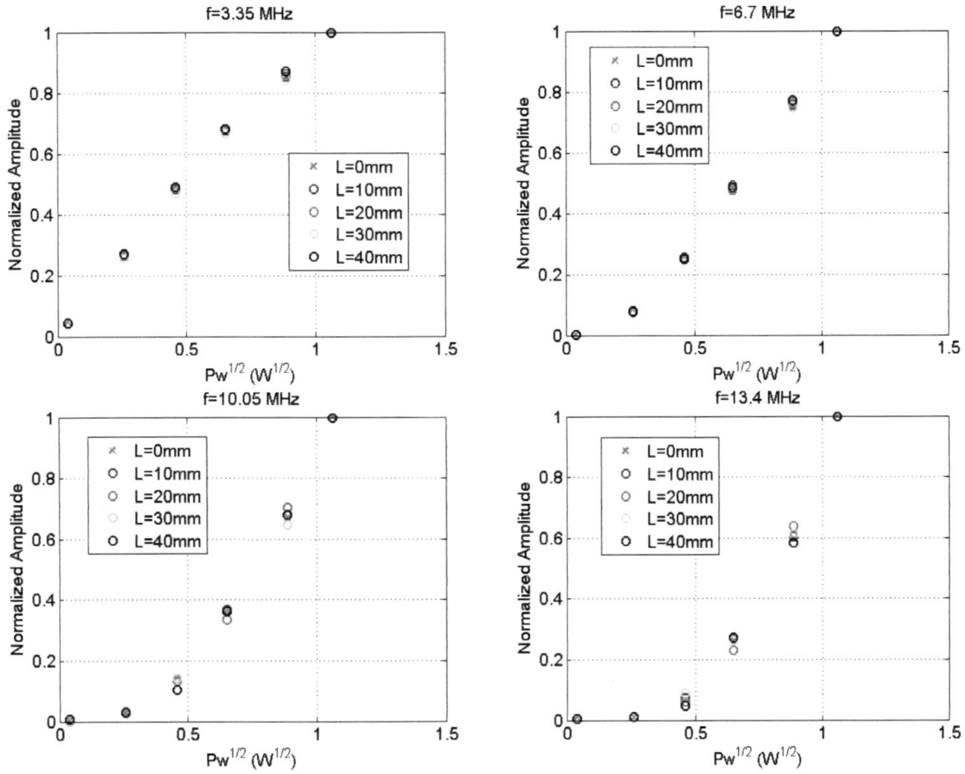

Figure 4. Harmonic amplitudes as a function of the square root of the electric power transmitted to the transducer, for different thicknesses L.

all reproduce the resonance at the frequency 2.8 MHz. Over the upper frequency interval (5.8-7.8 MHz) corresponding to the second harmonic, the estimate of the elastic mean free path is done with three sets of data corresponding to emission powers of 0.21, 0.79 and 1.1 W. Again, the agreement is good. Around the third harmonic (9.2-11.2 MHz), the elastic mean free path is determined with the two sets of data corresponding to emission powers of 0.79 and 1.1 W. The two estimates overlap. At last, with the set of data corresponding to the highest emission power (1.1 W), we could also estimate the elastic mean free path l_e over the frequency interval 12.6-14.6 MHz. The comparison with another set of data is not possible because this harmonic is not enough generated at lower emission amplitude.

4. Conclusion

Experimental measurements of the coherent field resulting from the interaction of a shock wave with a multiple scattering medium is reported. It shows that the transmitted coherent wave remains a shock wave despite high-order multiple scattering. The stiffening of the transmitted coherent waveform generates, in the frequency domain, higher order harmonics. Surprisingly, their amplitudes behave the same as in water, whatever the thickness of the scattering slab. Besides, estimates of the elastic mean free path l_e do not change with the emission power. It seems that the only effect associated with the shock wave is that it allows to estimate the elastic mean free path l_e on a frequency band whose extension increases with the emission power. Therefore, we can conclude that in our experimental configuration, multiple scattering does not act nonlinearly.

Diffuse acoustic wave spectroscopy makes use of the high sensitiveness of long time scattered

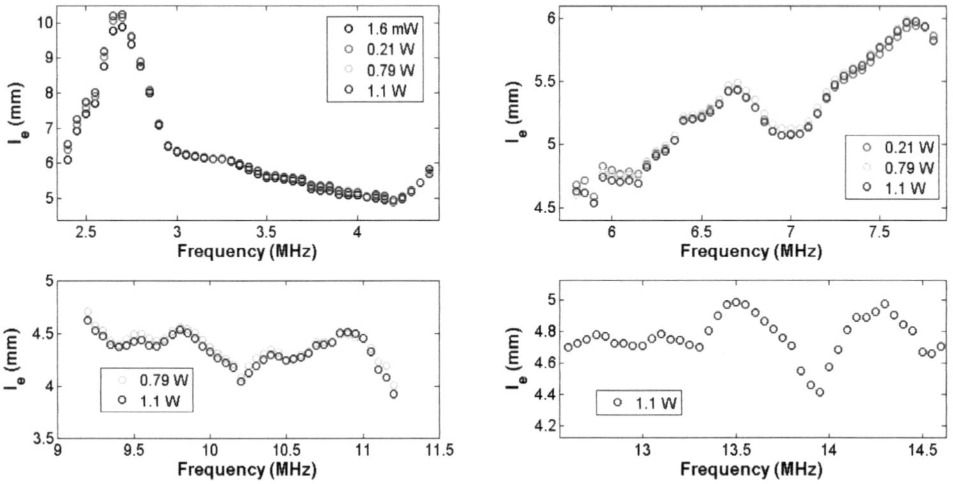

Figure 5. Frequency resolved estimate of the elastic mean free path l_e, for different emission powers. At the lowest emission power (1.6 mW), estimate of the elastic mean free path is only possible around the fondamental -first harmonic- whereas at the highest emission power (1.1 W), estimate of the elastic mean free path is made over 4 frequency bands corresponding to harmonics one to four.

signals (i.e., coda waves) to monitor small changes in a scattering medium [7, 8, 9, 10]. Future work will focus on these signals for which the cumulative nonlinear effects should be stronger.

Acknowledgements
The authors are grateful for funding provided by the Agence Nationale de la Recherche (ANR-11-BS09-007-01), for our four-year Research Project DiAMAN (2012-2015).

References

[1] Press A (ed) 1997 *Nonlinear Acoustics* (Hamilton, M. F. and Blackstock, D. T.)
[2] Markham J J, Beyer R T and Lindsay R B 1951 *Reviews of Modern Physics* **23** 353
[3] Averkiou M A and Hamilton M F 1997 *The Journal of the Acoustical Society of America* **102** 2539–2548
[4] Baker A C and Humphrey V F 1992 *The Journal of the Acoustical Society of America* **92** 1699–1705
[5] Baker A C, Anastasiadis K and Humphrey V F 1988 *The Journal of the Acoustical Society of America* **84** 1483–1487
[6] Derode A, Mamou V and Tourin A 2006 *Phys. Rev. E* **74** 036606
[7] Leroy V and Derode A 2008 *Phys. Rev. E* **77** 036602
[8] Cowan M L, Jones I P, Page J H and Weitz D A 2002 *Phys. Rev. E* **65** 066605
[9] Page J H, Cowan M L, Sheng P and Weitz D A 2001 *Diffusing acoustic wave transport and spectroscopy* (Kluwer Academic)
[10] Page J H, Cowan M L and Weitz D A 2000 *Physica B: Condensed Matter* **279** 130

Determination of the effective mechanical properties of inclusionary materials using bulk elastic waves

G Lepert[1,2], C Aristégui[1], O Poncelet[1], T Brunet[1], C Audoly[3] and P Parneix[2]

[a] *University of Bordeaux, CNRS, UMR 5295, I2M – Dpt. APY, 351, cours de la Libération, 33405 Talence*
[b] *DCNS Research, CEMIS, 83076 Toulon, France*
[c] *DCNS Research, CESMAN, 44620 La Montagne, France*
E-mail: guillaume.lepert@u-bordeaux1.fr
christophe.aristegui@u-bordeaux1.fr
patrick.parneix@dcnsgroup.com
christian.audoly@dcnsgroup.com

Abstract. For many years, the study of material with locally-resonant inclusions has known a great interest. Due to their attractive macroscopic acoustic properties, which do not exist in natural materials, many applications governed by the contrast between the constituent properties can be achieved: enhanced absorption, flat perfect lens, wave trapping… Such structures containing inclusions are usually modelled as effective homogeneous materials with complex-valued and frequency-dependent effective wave number and mass density. In this context, we derived a method for the simultaneous determination of both parameters using an immersed sample. First, its reliability is demonstrated using noisy synthesised temporal-waveform. Finally, the characterisation method is applied to experimental waveforms measured on a standard glass panel.

1 Introduction

Materials with inclusions are studied for many years in acoustics [1-5]. This kind of heterogeneous materials exhibits interesting acoustical macroscopic properties, such as strong absorption, band gap or negative index, for which the dynamical mass density and stiffness can take negative values.

An inclusionary medium can be considered as a homogeneous medium with the same acoustic behaviour than the heterogeneous one. For a viscoelastic homogeneous medium, it is known that its stiffness is complex valued and frequency dependent, in order to model the visco-thermal losses. Now, if we consider an inclusionary medium, its effective stiffness is also complex valued and frequency dependent in order to model the acoustic damping due to multiple scattering, in addition with eventual viscosity losses. Moreover, its mass density is also complex valued and frequency dependent, in order to take into account the inertial effects in between the different phases occurring when the wave propagates in the medium.

Many ultrasonic methods are proposed in the literature for characterising the phase celerity and attenuation of waves propagating in a homogeneous material. Here, in order to characterise materials with inclusions, we need a method that allows the identification of both the wave number and the dynamical mass density at the same time. Unfortunately, only few existing techniques retrieve the dynamical mass density. Adamowski [6] developed a characterisation method for the mass density of liquids, using reflected signals at different interfaces. Meanwhile, He [7] recovered the phase celerity and attenuation of a material using reflected and transmitted waveforms by a plate. These two techniques cannot be used to characterise materials with inclusions, because the mass density is assumed to be real valued and constant with the angular frequency ω. In addition, since we do not want to be subjected to any *a priori* model describing the variation of the material properties with the frequency, the characterisation procedure is performed "frequency by frequency".

An identification procedure using reflection and transmission of an immersed plate is developed here. This method for the retrieval of the wave number and complex-valued dynamical mass density is based on that proposed by Fokin [8]. It is suitable for the experimental setup presented in section 2. Here, the technique is applied on temporal waveforms synthesised from the acoustic response of a viscoelastic homogeneous immersed plate. Its reliability on a simple case will ensure a part of reliability on other cases. More, the sensitivity of the technique regarding the measurement uncertainties is investigated.

2 Identification method

2.1 Experimental setup

The identification method presented here is based on the measurement of the reflection and transmission coefficients of a plate. The experimental setup contains a water tank within which a plate shape material is immersed. Two transducers are placed on both sides of the sample, and thus are able to catch the signals reflected and transmitted by the material (Fig. 1). The transducers send a temporal signal with a finite temporal length and a large frequency-band, centred at the central frequency of the transducers. Six signals can be measured by the setup, as it is shown on figure 1.

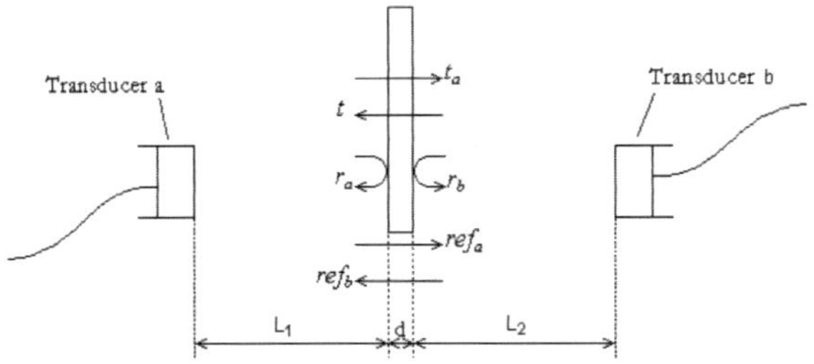

Figure 1: scheme of the experimental setup, and all the measurable signals.

Depending on the plate material, the overall shape of the reflected and transmitted signals can be slightly different (Fig. 2). Indeed, two general cases can be observed. First (Fig. 2a), the echoes coming from the interfaces can be separated if the thickness is large enough, or if the absorption is small enough. In fact, more the plate thick is, more the shift between the echoes large is, so it will be easier to separate the different echoes. It is interesting to note that, whatever the attenuation level is, the successive echoes have decreasing amplitude, until the total absorption of the signal. Second (Fig. 2b and 2c), the echoes cannot be separated, if the attenuation is large enough (Fig. 2b), or if the thickness is small enough (Fig. 2c). For thin plates, all the echoes are contained in the main signal. Their number is *a priori* unknown, since it is large for weak absorption or equal to zero for weak absorption. In the case of strong absorption, if no acoustic signal is transmitted by the plate, the identification method is not applicable, since the knowledge of both the reflection and transmission spectra at the same time is required. In such case, the measurement of the only reflection spectrum does not allow the two-parameter recovery.

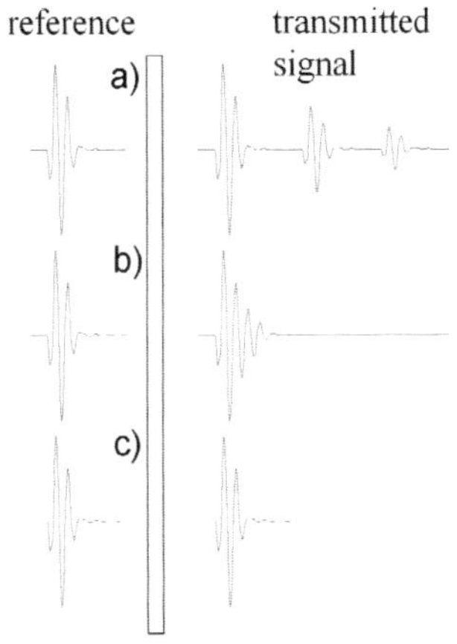

Figure 2: different configurations depending on the material properties / thickness values: a) the echoes are separable (thick plate); b) all echoes are contained in the main signal (thin plate); c) only the direct signal is present, because of the strong absorption in the plate (no back and forth reflections within the plate)

In this paper, the subscript 0 refers to the external medium (water), the subscript 1 refers to the plate material, and c is the longitudinal wave celerity. The properties we will identify are the wave number $k(\omega)=\omega/c(\omega)+\mathrm{i}\alpha(\omega)$, with α standing for the attenuation, and the complex-valued dynamical mass density $\rho(\omega)=\rho'(\omega)+\mathrm{i}\rho''(\omega)$, with $(\rho',\rho'')\in\mathbb{R}^2$.

Let t_a, t_b, be the signals transmitted through the plate respectively from transducer a to transducer b, and from transducer b to transducer a. The signals r_a and r_b are, in the same way, the signals reflected by the plate. Finally, ref_a and ref_b are the reference signals, only propagating through the water without the sample.

2.2 Characterisation method

2.2.1 Identification using total reflection and total transmission

Let T_a, T_b, R_a, R_b, Ref_a et Ref_b be the spectra coming from the temporal-waveforms recorded by the experimental setup. These spectra can be expressed using the reflection and transmission coefficients of the plate, respectively $R(\omega)$ and $T(\omega)$, as well as the transfer functions of the transducers ($X_a(\omega)$, $X_b(\omega)$ in emission, and $Y_a(\omega)$, $Y_b(\omega)$ in reception):

$$\begin{aligned}
T_a(\omega) &= X_a(\omega)\mathrm{e}^{\mathrm{i}k_0L_1}T(\omega)\mathrm{e}^{\mathrm{i}k_0L_2}Y_b(\omega),\\
T_b(\omega) &= X_b(\omega)\mathrm{e}^{\mathrm{i}k_0L_2}T(\omega)\mathrm{e}^{\mathrm{i}k_0L_1}Y_a(\omega),\\
R_a(\omega) &= X_a(\omega)\mathrm{e}^{\mathrm{i}k_0L_1}R(\omega)\mathrm{e}^{\mathrm{i}k_0L_1}Y_a(\omega),\\
R_b(\omega) &= X_b(\omega)\mathrm{e}^{\mathrm{i}k_0L_2}R(\omega)\mathrm{e}^{\mathrm{i}k_0L_2}Y_b(\omega),\\
Ref_a(\omega) &= X_a(\omega)\mathrm{e}^{\mathrm{i}k_0(L_1+d+L_2)}Y_b(\omega),\\
Ref_b(\omega) &= X_b(\omega)\mathrm{e}^{\mathrm{i}k_0(L_2+d+L_1)}Y_a(\omega),
\end{aligned} \tag{1}$$

with $k_0=\omega/c_0$ and where the term $\mathrm{e}^{-i\omega t}$ has been omitted. In the system (1), we suppose that the R and T functions are the same in both propagation paths. This assumption allows us to extract the coefficients R and T:

$$\frac{T_a}{Ref_a} = \frac{T_b}{Ref_b} = T\mathrm{e}^{-\mathrm{i}k_0 d}, \qquad \frac{R_a}{Ref_a}\frac{R_b}{Ref_b} = R^2\mathrm{e}^{-2\mathrm{i}k_0 d}. \tag{2}$$

These combinations of spectra eliminate the unknown transfer functions of the transducers, as well as the distances between the plate and the transducers. More, this measurement procedure of signals avoids the knowledge of a reference signal usually needed for the measurement in reflection.

The analytical expressions for R and T are functions of the plate properties [8]:

$$R = \frac{\left(\frac{1}{Z} - Z\right)\tan(k_1 d)\mathrm{i}}{2 - \left(\frac{1}{Z} + Z\right)\tan(k_1 d)\mathrm{i}}, \qquad T = \frac{2}{\cos(k_1 d)\left[2 - \left(\frac{1}{Z} + Z\right)\tan(k_1 d)\mathrm{i}\right]}, \tag{3}$$

with $Z = \frac{Z_1}{Z_0} = \frac{\rho_1 k_0}{\rho_0 k_1}$.

Using the experimental measurement of these coefficients, an identification procedure of the material parameters is developed here. It consists, for a given frequency, to invert analytically the system (3), and then to extract analytical expressions for the wave number k_1 and the mass density ρ_1:

$$\cos(k_1 d) = \frac{1 - R^2 + T^2}{2T} = A,$$

and then

$$k_1 = \frac{\pm\cos^{-1}(A) + 2\pi n}{d}, (n \in \mathbb{Z}) \tag{4}$$

$$\left(\frac{1}{Z} + Z\right) = \left(2 - \frac{2}{T\cos(k_1 d)}\right)\frac{1}{\mathrm{i}\tan(k_1 d)} = B,$$

and then

$$Z = \frac{B \pm \sqrt{B^2 - 4}}{2}. \tag{5}$$

The wave number k_I is defined with sign uncertainty as well as branch selection incertitude (with n). The sign of the wave number is set for its imaginary part to be positive. Indeed, a negative imaginary part corresponds to an amplification of the wave during its propagation through the material (non passive material). We observe that the acoustic impedance Z is also defined with a sign uncertainty (Eq. 5).

2.2.2 Identification using partial reflection and partial transmission

When the different echoes in a signal are separable, the sought parameters can be identified using the direct reflection R_1, and the direct transmission T_1, without any multiple reflections within the plate. These coefficients can be expressed, like R and T, using the properties of the material:

$$R_1 = \frac{1-Z}{1+Z}, \quad T_1 = \frac{4Z}{(1-Z)^2} e^{ik_1 d}. \tag{6}$$

As previously, we can analytically revert this system:

$$e^{ik_1 d} = \frac{T_1}{1-R_1^2} = C,$$

and then
$$k_1 = \frac{-i\ln(C) + 2\pi n}{d}, (n \in \mathbb{Z}) \tag{7}$$

$$\frac{Z}{(1+Z)^2} = \frac{T_1}{4} e^{-ik_1 d} = D,$$

and then
$$Z = \frac{1-2D \pm \sqrt{1-4D}}{2D}. \tag{8}$$

3 Stability of the characterisation

In order to quantify the characterisation stability, we simulate the experimental setup with synthetic data. The material is chosen to be linear viscoelastic $(c_1(\omega) \in \mathbb{C}, \rho \in \mathbb{R}^+)$, and to respect causal relations between phase celerity and attenuation [9, 10] (Fig. 3). This causal mechanical system is based on relaxation functions, which consider appropriate probability densities of relaxation times. These judicious choices lead one to explicit relations between the mechanical characteristics of homogeneous linear viscoelastic media and observable acoustic properties (wave number).

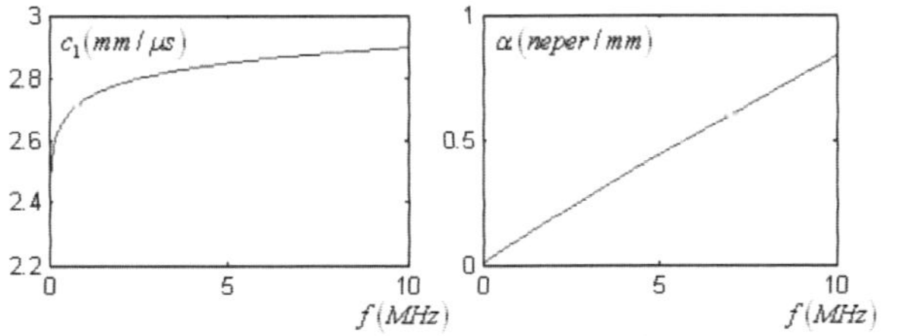

Figure 3: phase celerity and attenuation of the material respecting causality. $(\rho_1 = 1.56 g/cm^3)$.

The reflected and transmitted signals are simulated using the coefficients *R* and *T*. The reference signal is an experimental signal, measured with two transducers with a central frequency of 150kHz (Fig. 4).

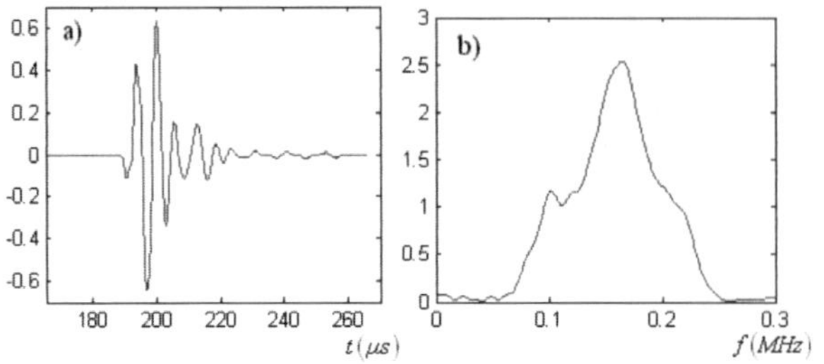

Figure 4: a) reference signal (au); b) modulus of the spectra of the reference signal (au)

The characterisation procedure applied to synthetic signals gives exactly the original data used for the simulation. The goal is then to observe the reliability of the identification for noisy signals. The white noise is expressed in dB with a signal to noise ratio (SNR), defined as the energy ratio between the original signal and the noise. We choose here a 24dB SNR. The signals represented in figures 5 and 6 have been simulated for two thicknesses (10mm and 1mm). The results of the identification are shown in figures 7 and 8.

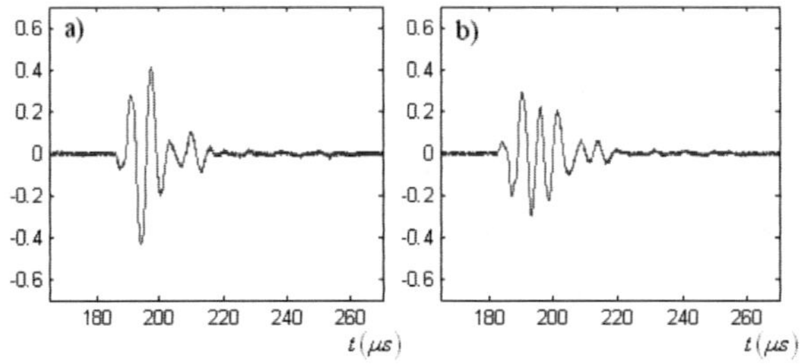

Figure 5: temporal waveform simulated in transmission a) and in reflection b) for a 10mm-thick plate, with a 24dB SNR.

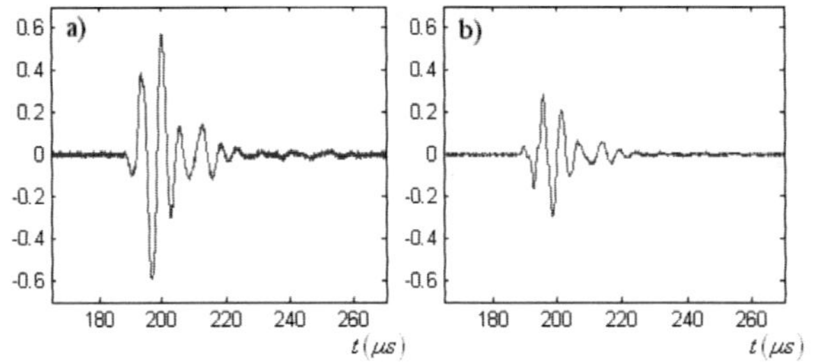

Figure 6: temporal waveform simulated in transmission a) and in reflection b) for a 1mm-thick plate, with a 24dB SNR.

We can observe that, with the same noise level, the identification of the wave number is less accurate for a thin plate. Concerning the mass density, the identification has the same sensitivity as that of the wave number for the 10mm-thick plate, and its sensitivity is of the same level as that of the 1mm-thick plate. This result is directly linked to the ratio of the wave length regarding the plate thickness. Indeed, if the wave length is much larger than the plate thickness, the propagation inside the plate is very short, so the acoustic response of the thin plate is less dependent upon the wave number than for the thick plate. At the central frequency of the reference signal (160kHz), the wave length is equal to 16mm, which is much larger than the thickness of the thinnest plate (1mm).

It seems that a thickness comparable to the wave length of the signal is required to have an accurate retrieval of the dynamic parameters.

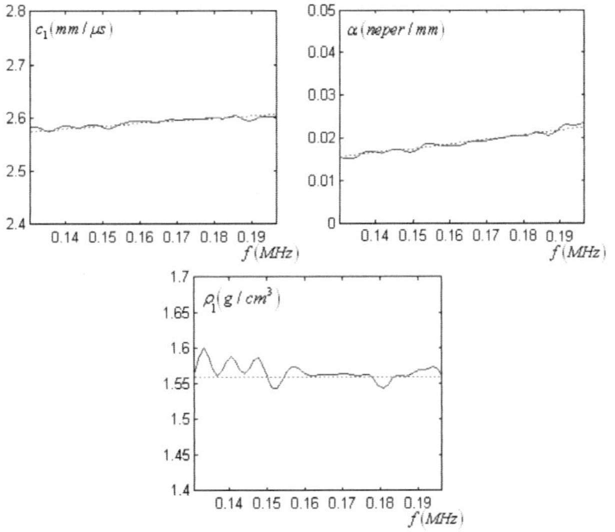

Figure 7: phase celerity, attenuation and mass density (retrieved properties in blue, original data in green), recovered using signals simulated for a 10mm-thick plate with a 24dB SNR.

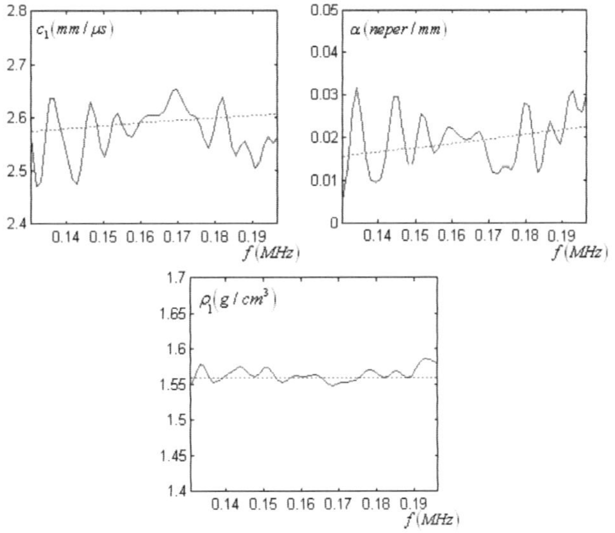

Figure 8: phase celerity, attenuation and mass density (retrieved properties in blue, original data in green), recovered using signals simulated for a 1mm-thick plate with a 24dB SNR.

4 Experimental results

The reflection and transmission coefficients have been measured with a 7.97mm-thick glass plate (with a pair of 5MHz central-frequency transducers).

Figure 9 shows the results of the characterisation method applied on the experimental waveforms. Due to the high central frequency of the transducers, the direct response of the plate (without multiple reflections inside the plate) can be extracted from the signals. While the retrieved phase celerity corresponds to values found in literature, the attenuation is higher than expected. This overestimating can be explained by the fact that the experimental setup is not fully modelled. Indeed, an erroneous alignment of the transducers reduces the amplitude of the received signal, and so increases the recovered attenuation. More, the diffraction of the acoustic field, due to the finite size of the transducers also reduces the received signal amplitude.

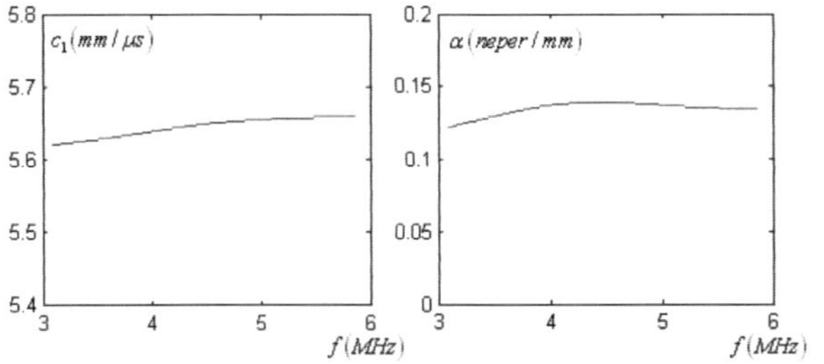

Figure 9: phase celerity and attenuation recovered from the measured reflection and transmission of a 7.97mm thick glass plate.

The retrieved mass density is represented in figure 10. As a function of the sign we choose for the acoustic impedance (Eq. 5), two values can be proposed for the mass density. The real part of the first value of the identified mass density (in blue in figure 10) corresponds to the real density of glass (2.51g/cm^3), but its imaginary part is higher than expected (since the attenuation is also overestimated). The second solution for the mass density is very low, for its real part as well as for its imaginary part and does not correspond with the expected one.

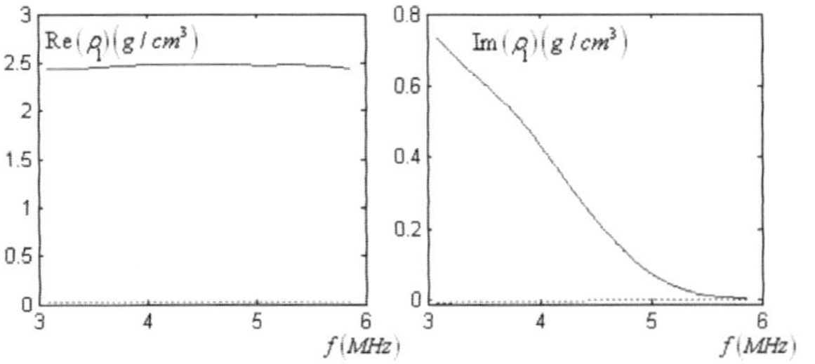

Figure 10: two values for the mass density (real and imaginary parts) recovered from the measured reflection and transmission of a 7.97mm-thick glass plate (first value in solid blue line, and second value in dotted green line).

5 Conclusions

An ultrasonic characterisation technique for *longitudinal* wave properties of inclusionary materials has been developed. The method retrieves both the effective wave number and the effective mass density from the reflection and transmission of an immersed plate. Since no model for the evolution of the parameters is taken into account, explicit formulas for these parameters have been derived depending on the frequency.

We then studied numerically the sensitivity study of the method to the noise in waveforms. The characterisation procedure is found to be more sensitive to the noise when the plate is thin (especially for the wave number)

Finally, the experimental characterisation of a glass plate has shown that the current procedure naturally overestimates the attenuation.

Acknowledgements

The authors would like to thank particularly Benoît Mascaro, who participates in the experimental aspect of this work. Guillaume Lepert is indebted to the *Direction Générale de l'Armement* for its financial support.

References

[1] Kuster G T and Toksöz M N 1974, Velocity and attenuation of seismic waves in two-phase media: Part I. Theoretical formulations, *Geophysics* **39** (5), p. 587-606

[2] Kinra V K, Petraitis M S and Datta S K 1979, Ultrasonic wave propagation in a random particulate composite, *Int. J. Sol. Struct.* **16**, p. 301-312

[3] Waterman P C and Truell R 1961, Multiple scattering of waves, *J. Math. Phy.* **2** (4), p. 512

[4] Aristégui C and Angel Y C 2007, Effective mass density and stiffness derived from P-wave multiple scattering, *Wave Motion* **44**, p. 153-164

[5] Audoly C 1994, Acoustic analysis of panels made with viscoelastic materials containing resonant cavities, *Acta acustica* **2**, p. 393-402

[6] Adamowski J C, Buiochi F, Simon C and Silva E C N 1995, Ultrasonic measurement of density of liquids, *J. Acoust. Soc. Am.* **97** (1), p. 354-361

[7] He P 2000, Measurement of acoustic dispersion using both transmitted and reflected pulses, *J. Acoust. Soc. Am.* **107** (2), p. 801-807

[8] Fokin V, Ambati M, Sun C and Zhang X 2007, Method for retrieving effective properties of locally resonant acoustic metamaterials, *Physical review B* **76**, 144302

[9] Wintle H J 1999, Kramers-Kronig analysis of polymer acoustic data, *J. App. Phy.* **85** (1), p. 44-48

[10] Wintle H J 2000, Power law absorption in polymers and other systems, *J. Acoust. Soc. Am.* **107** (3), p. 1770-1773

Simulation of ultrasonic wave propagation in welds using ray-based methods

A. Gardahaut[1], K. Jezzine[1], D. Cassereau[2,3] and N. Leymarie[1]
[1]CEA, LIST, F91191 Gif-Sur-Yvette, France
[2]CNRS, UMR 7623, LIP, 15 rue de l'école de médecine, 75006 Paris, France
[3]ESPCI ParisTech, 10 rue Vauquelin, 75005 Paris, France
E-mail: audrey.gardahaut@cea.fr

Abstract. Austenitic or bimetallic welds are particularly difficult to control due to their anisotropic and inhomogeneous properties. In this paper, we present a ray-based method to simulate the propagation of ultrasonic waves in such structures, taking into account their internal properties. This method is applied on a smooth representation of the orientation of the grain in the weld. The propagation model consists in solving the eikonal and transport equations in an inhomogeneous anisotropic medium. Simulation results are presented and compared to finite elements for a distribution of grain orientation expressed in a closed-form.

1. Introduction

Ultrasonic Non Destructive Testing (NDT) is a major technique used in various industrial sectors such as nuclear, aerospace, petrochemical for the inspection of welds. Because of their polycrystalline structure, the detection and characterization of defects located inside or in the vicinity of welds can be complicated. Some disturbances of the beam such as splitting and skewing can be observed on experimental structural echoes [1]. In this context, the simulation of ultrasonic inspection is helpful to have a better understanding of the physical phenomena in complex media. Numerical methods such as finite element [2] may be used to simulate the propagation of ultrasonic waves in welds. Alternatively, a semi-analytic propagation model based on ray theory has been implemented in the CIVA software [3] [4]. This model has to be applied on a piecewise description of the weld. A set of several homogeneous domains with a constant crystallographic orientation is considered. In this case, the rays propagate along a constant direction of energy for a given slowness vector and the reflection and refraction are calculated at each interface. In this paper, we propose to extend the ray model implemented in CIVA to take account of a continuously variable description of the grain orientation in the weld.

2. Context: Ultrasonic simulation of wave propagation in welds

A weld is an anisotropic and inhomogeneous medium with an internal grain organization that depends on the steel shade and the parameters of the welding process such as the velocity, the electrode diameter or the welding technique. The difficulties of on-site inspection are due to this polycrystalline structure possibly responsible for beam splitting and skewing. The Dynamic Ray Tracing method (DRT) usually applied in Geophysics [5] will be used to simulate the ultrasonic propagation in welds. The geometry of the weld, the elasticity constants and the attenuation

Journal of Physics: Conference Series **498** (2014) 012008 doi:10.1088/1742-6596/498/1/012008

of the welding materials and the crystallographic orientation of the grain at any position in the weld are required as input data for DRT. These data may be obtained either in a closed-form [6] or thanks to an image processing technique [7] applied on the macrograph as shown in Figure 1. It is important to note that, as the DRT relies on a high frequency approximation, it has to be applied on a slowly continuously varying description of the orientation of the weld.

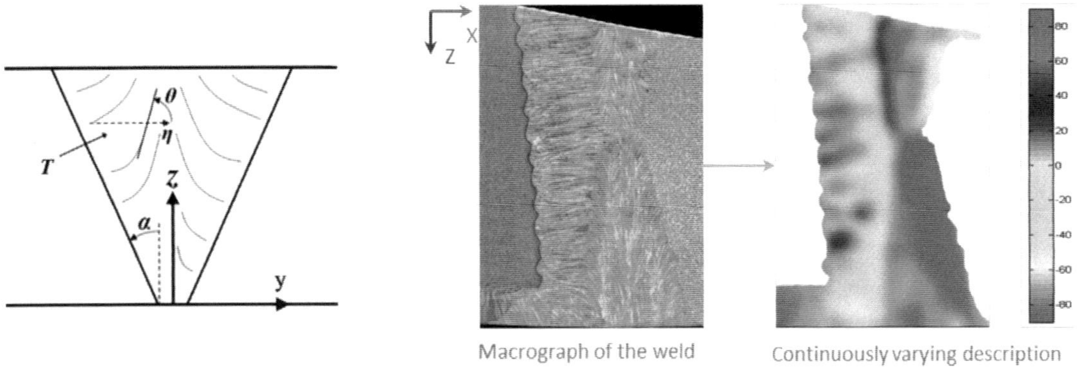

(a) (b)

Figure 1. Possible descriptions of the crystallographic orientation of a weld: (a) Closed-form expression proposed by Ogilvy [6] for a V-shaped weld, (b) Description obtained from a macrograph for a bimetallic weld [7].

To evaluate the ray-paths and the travel time, we have to solve the *eikonal equation* (1).

$$\left(\nabla T^2\right) = \frac{1}{c(x)^2} \tag{1}$$

The eikonal equation represents a non linear partial differential equation of the first order for the travel time $T(x)$. It describes the propagation of the wavefront $T(x)$ of the considered wave and depends on the phase velocity $c(x)$ of this wave.

A second equation called the *transport equation* (2) represents a linear partial differential equation of the first order in $A(x)$, the amplitude function. It is solved along a ray Ω and describes the conservation of the energy inside a ray tube.

$$A(x)\nabla^2 T(x) + 2\nabla A(x).\nabla T(x) = 0 \tag{2}$$

3. Dynamic Ray Tracing Model for smooth description of weld

3.1. Ray theory

To obtain the ray trajectories, a differential ray tracing system is derived from the eikonal equation [5] and written in Hamiltonian form:

$$\left\{ \begin{array}{lcl} \dfrac{dx_i}{dT} & = & a_{ijkl} p_l g_j^{(m)} g_k^{(m)} = V_i^e, \\ \dfrac{dp_i}{dT} & = & -\dfrac{1}{2}\dfrac{\partial a_{ijkl}}{\partial x_i} p_k p_n g_j^{(m)} g_l^{(m)}, \end{array} \right. \tag{3}$$

where T is the travel time, a_{ijkl} the elasticity constants of the medium at position x_i normalized by the density ρ. p_i represents the components of the slowness vector, $g_j^{(m)}$ the eigenvectors of the Christoffel tensor [5] are the components of the polarization vector and V_i^e is the energy velocity. This system called axial ray system is composed of two coupled ordinary differential equations. The first one describes the variations of the position with respect to the travel time and the second one the variations of the slowness.

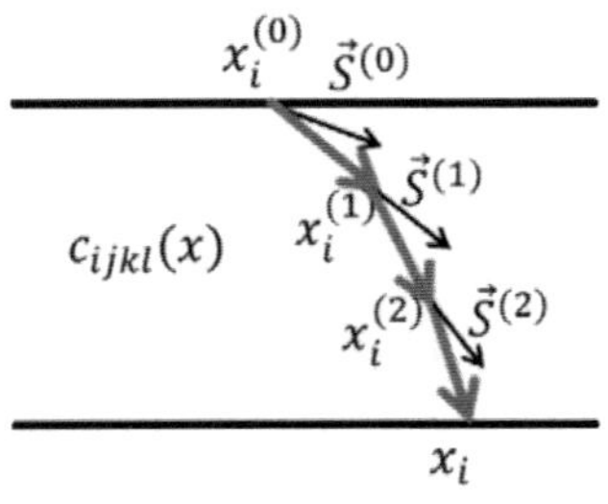

Figure 2. Principle of an iterative solving of a ray system.

In order to obtain the trajectories of the ray, the axial ray system has to be solved. Taking into account the initial conditions of the system, we can use classic iterative techniques (Figure 2) such as the Euler or Runge-Kutta methods [8].

The amplitude of the ray is evaluated by solving the transport equation (2) in inhomogeneous anisotropic media. By deriving the axial ray system with respect to an initial parameter γ, we obtain a system of linear ordinary differential equations of the first order for $Q_i^{(x)}$ and $P_i^{(x)}$ (4). These quantities represent respectively the first partial derivatives of x_i and p_i with respect to γ. The parameter γ may represent any parameter of the ray Ω. For example, it can be chosen as the take-off angle between the axial ray and a paraxial one.

$$\begin{cases} \dfrac{d}{dT}\left(\dfrac{dx_i}{d\gamma}\right) = \dfrac{dQ_i^{(x)}}{dT} = \dfrac{1}{2}\dfrac{\partial^2 G^{(x)}}{\partial p_i^{(x)}\partial\gamma} = A_{ij}^{(x)}Q_j^{(x)} + B_{ij}^{(x)}P_j^{(x)}, \\ \dfrac{d}{dT}\left(\dfrac{dp_i^{(x)}}{d\gamma}\right) = \dfrac{dP_i^{(x)}}{dT} = -\dfrac{1}{2}\dfrac{\partial^2 G^{(x)}}{\partial x_i \partial\gamma} = -C_{ij}^{(x)}Q_j^{(x)} - D_{ij}^{(x)}P_j^{(x)}. \end{cases} \tag{4}$$

G represents any of the three eigenvalues of the Christoffel tensor and is expressed as $G_m^{(x)} = a_{ijkl}p_j^{(x)}p_l^{(x)}g_i^{(x)}g_k^{(x)}$. Three eigenvalues are evaluated, associated to three eigenvectors $g_i^{(m)}$ of the Christoffel tensor representing the three plane waves that propagate in the medium.

The paraxial ray system is solved simultaneously with the axial ray system by using a numerical iterative technique. This paraxial ray system detailed in [5] can be written with a matrix formulation (5). The equation (4) is the paraxial ray system expressed in a Cartesian regular coordinates system x_i where the indexes i and j are equal to 1, 2 or 3. Indeed the paraxial ray system in this coordinate system consists of six scalar linear ordinary differential equations. It may be more convenient to express the paraxial ray system in a coordinate system that allows to decrease the number of equations. The expression (5) also represents the paraxial ray system in a wavefront orthonormal coordinates system y_i where the indexes M and N are equal to 1 or 2. In this coordinate system, the origin moves along the central ray with the propagating wavefront. The y_3-axis is oriented along the slowness vector $\vec{p}$ at the origin point, and axes y_1 and y_2 are in the plane tangent to the wavefront at the origin point and are mutually perpendicular. The system consists then of a four scalar linear differential equations.

$$\begin{pmatrix} Q_M^{(y)(r+1)} \\ P_M^{(y)(r+1)} \end{pmatrix} = \begin{pmatrix} \Pi_{11}^{(y)} & \Pi_{12}^{(y)} \\ \Pi_{21}^{(y)} & \Pi_{22}^{(y)} \end{pmatrix} \cdot \begin{pmatrix} Q_N^{(y)(r)} \\ P_N^{(y)(r)} \end{pmatrix} = \begin{pmatrix} 1 + A_{MN}^{(y)}\Delta T & B_{MN}^{(y)}\Delta T \\ -C_{MN}^{(y)}\Delta T & 1 - D_{MN}^{(y)}\Delta T \end{pmatrix} \cdot \begin{pmatrix} Q_N^{(y)(r)} \\ P_N^{(y)(r)} \end{pmatrix}. \tag{5}$$

Mathematically, a paraxial ray of the tube is described as a 4x1 vector composed of the paraxial quantities [5]. Q_1 and Q_2 represent the spatial deviation of the paraxial ray from the axial ray respectively in the directions x_1 and x_2 and P_1 and P_2 the slowness deviation of the paraxial ray from the axial one in the same directions as shown in Figure (3).

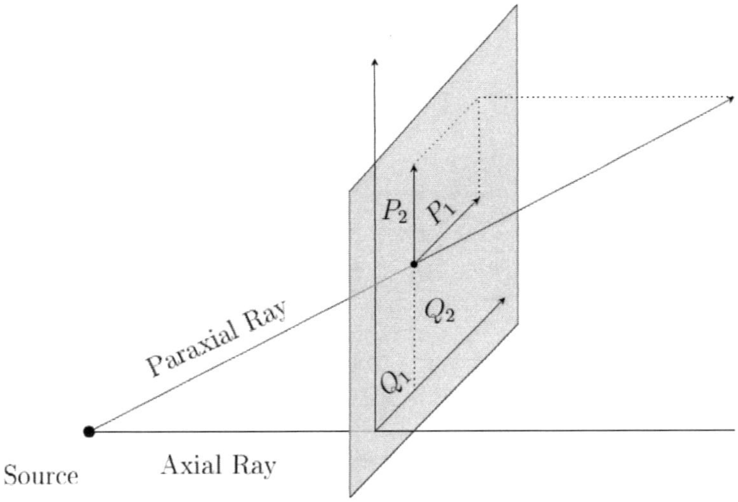

Figure 3. Representation of the quantities describing a paraxial ray.

In the dynamic ray tracing model, the paraxial quantities Q_i and P_i at iteration $(r + 1)$ depend on the same quantities evaluated at the previous position of calculation (r) through a 4x4 propagation matrix. At the end of the propagation, we obtain:

$$\begin{pmatrix} Q_M^{(y)(r+1)} \\ P_M^{(y)(r+1)} \end{pmatrix} = L_{(r)}^{(y)} \cdot L_{(r-1)}^{(y)} \dots L_{(1)}^{(y)} \cdot L_{(0)}^{(y)} \cdot \begin{pmatrix} Q_N^{(y)(0)} \\ P_N^{(y)(0)} \end{pmatrix} = \begin{pmatrix} \Pi_{11_{tot}}^{(y)} & \Pi_{12_{tot}}^{(y)} \\ \Pi_{21_{tot}}^{(y)} & \Pi_{22_{tot}}^{(y)} \end{pmatrix} \cdot \begin{pmatrix} Q_N^{(y)(0)} \\ P_N^{(y)(0)} \end{pmatrix}. \tag{6}$$

At each time step, the eikonal and transport equations are solved. This means that the matrices $A_{MN}^{(y)}$, $B_{MN}^{(y)}$, $C_{MN}^{(y)}$ and $D_{MN}^{(y)}$ are re-evaluated at each step. The propagation matrix is then updated as shown in Figure 4 and the position and the slowness vector evaluated from the axial ray system.

Then, the geometrical spreading of the ray tube between the iterations (0) and $(r + 1)$ can be evaluated thanks to the $\Pi_{12_{tot}}$ matrix of the propagation matrix between the source and the receiver points (7):

$$\mathcal{L}(r+1,0) = |det\Pi_{12_{tot}}(r+1,0)|^{1/2}. \tag{7}$$

In order to apply this method, we have made an approximation that consists in considering the medium locally homogeneous between each time step. This means that the matrices $\Pi_{11_{tot}}$ and $\Pi_{22_{tot}}$ are equal to the identity matrix and the matrix $\Pi_{21_{tot}}$ is equal to zero at each time step. The matrix $\Pi_{12_{tot}}$ is the only one re-evaluated at each time step.

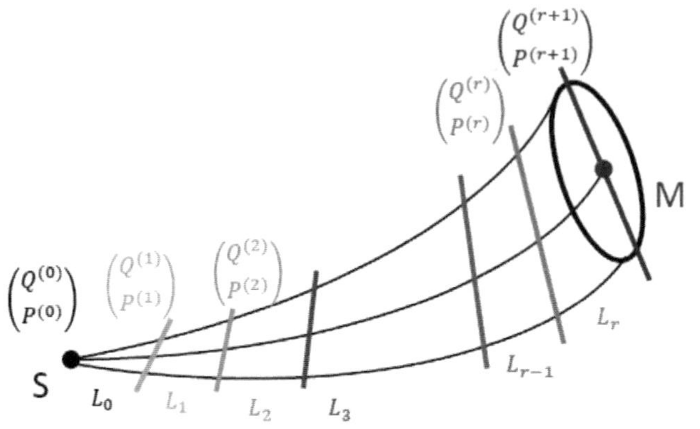

Figure 4. Representation of the evolution of a ray tube and the paraxial quantities Q_i and P_i during the propagation.

3.2. Application and validation of the model

We have applied the dynamic ray tracing model using a closed-form expression to describe the distribution of grain orientation proposed by Ogilvy [6]:

$$\theta = \begin{cases} \arctan\left(\dfrac{T(D + z * \tan\alpha)}{x^n}\right), & \text{for } x \geq 0, \\ -\arctan\left(\dfrac{T(D + z * \tan\alpha)}{(-x)^n}\right), & \text{for } x < 0. \end{cases} \tag{8}$$

This mathematical description is adapted for V-shaped weld composed of a transversely isotropic medium. Parameters D and α are geometrical: D is the half-width of the bottom of the weld and α is the angle between the z-axis and the beveled edge as shown in Figure 5(a). The quantity η describes the variation of the orientation in the x direction and T is proportional to the tangent of the grain axis at the beveled edge. The ray trajectories obtained with the dynamic ray tracing model have been compared to the results of Connolly's PhD thesis [9]. Connolly also used a ray-based model to simulate the propagation of ultrasonic waves in welds but at each time step, virtual interfaces are created so that the reflection and transmission coefficients can be evaluated. The parameters for (8) are: $D = 2$mm, $\alpha = 21.80°$, $T = 1$ and $\eta = 1$. In Figure 5(b) we can observe a very good agreement between both models for a ray propagating through the V-weld.

A hybrid finite element code developed by EDF and CEA is used as a validation tool [10] of the computed wave field. In this model, a finite element computation made by ATHENA is performed in a box containing the weld and takes an incident field computed by CIVA as input data. A comparison of the wave field is shown in Figure 6 for the same V-weld described with (8). A good agreement is observed on the wave field evaluated by both models. But we can observe slight discrepancies in the middle of the weld where the dynamic ray tracing model seems to overestimate the amplitude. We suppose that these differences between the two models can be explained by the approximation considering the medium locally homogeneous between each time step. Indeed, the paraxial quantities are not rigorously evaluated so the geometrical spreading is approximated.

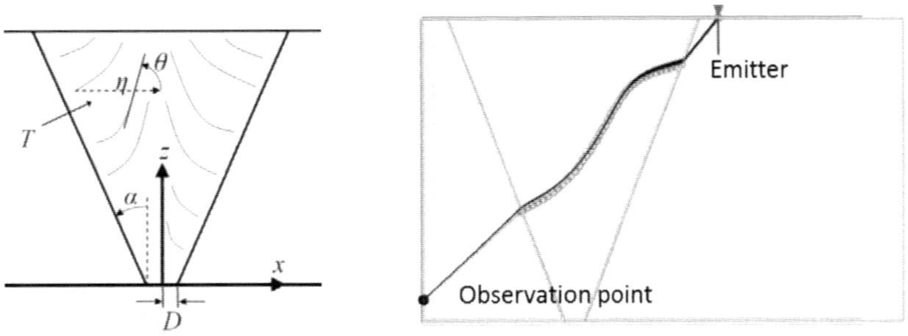

(a) (b)

Figure 5. (a) Parameters of a closed-form description of the crystallographic orientation of the weld proposed by Ogilvy [6]. (b) Representation of the ray trajectories in a bimetallic weld described by this law - (oo) Previously published results [9] and (–) Dynamic ray tracing results.

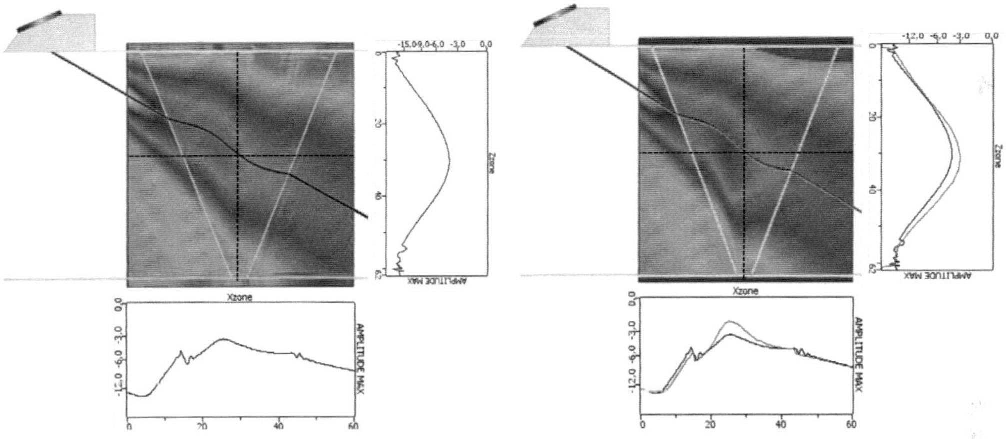

(a) (b)

Figure 6. Comparison of the maximum particle velocity field for L60° inspection at 2, 25 MHz obtained with (a) ( ) the dynamic ray tracing model implemented in the CIVA software and (b) (–) the hybrid finite element model (CIVA/ATHENA).

4. Conclusion and perspectives

A ray-based model has been applied on a continuously varying representation of the orientation of the grains in a weld. This method has been validated with the literature for a V-weld described by a closed-form expression. But some discrepancies most likely due to the approximated evaluation of the paraxial quantities and the geometrical spreading are observed in the computed wave field obtained with a hybrid finite element model and the ray-based model that has been developed.

In order to evaluate more precisely the wave field we are working on a rigorous computation of the paraxial quantities (5) and therefore the geometrical spreading. We intend to validate it numerically and experimentally on a realistic continuously varying description obtained from a macrograph of a weld thanks to an image processing technique.

References

[1] Ahmed F, Poidevin C and Serre M 26-29 May 1998 *Proceedings of the 7th European Conference on Non-Destructive Testing*
[2] Apfel A, Moysan J, Corneloup G, Fouquet T and Chassignole B 2005 *Ultrasonics 43, 447-456*
[3] CIVA software platform for simulating NDT techniques (UT, EC, RT) http://www-civa.cea.fr
[4] Lhémery A, Calmon P, Lecoeur-Taïbi I, Raillon R and Paradis L 2000 *NDT&E Int. 37, 499-513*
[5] Červený V 2001 *Seismic Ray Theory* (Cambridge University Press)
[6] Ogilvy J 1985 *NDT International 18, 67-77*
[7] Gardahaut A, Jezzine K and Cassereau D 2012 *Proceedings of the Acoustics 2012 Nantes Conference*
[8] Journiac S, Leymarie N, Dominguez N and Potel C 2011 *Journal of Physics: Conference Series 269*
[9] Connolly G 2009 *Modelling of the propagation of ultrasound through austenitic welds* Ph.D. thesis Imperial College of London
[10] Mahaut S, Leymarie N, Poidevin C, Fouquet T and Dupond O 2011 *Insight 53, 664-667*

ATHENA 3D: A finite element code for ultrasonic wave propagation

C Rose[1], F Rupin[2], T Fouquet[1], B Chassignole[2]
[1]EDF R&D SINETICS, Clamart (92), France
[2]EDF – R&D MMC, Sites des Renardières, Moret sur Loing (77) France

E-mail : Christian.rose@edf.fr

Abstract. The understanding of wave propagation phenomena requires use of robust numerical models. 3D finite element (FE) models are generally prohibitively time consuming. However, advances in computing processor speed and memory allow them to be more and more competitive. In this context, EDF R&D developed the 3D version of the well-validated FE code ATHENA2D. The code is dedicated to the simulation of wave propagation in all kinds of elastic media and in particular, heterogeneous and anisotropic materials like welds. It is based on solving elastodynamic equations in the calculation zone expressed in terms of stress and particle velocities. The particularity of the code relies on the fact that the discretization of the calculation domain uses a Cartesian regular 3D mesh while the defect of complex geometry can be described using a separate (2D) mesh using the fictitious domains method. This allows combining the rapidity of regular meshes computation with the capability of modelling arbitrary shaped defects. Furthermore, the calculation domain is discretized with a quasi-explicit time evolution scheme. Thereby only local linear systems of small size have to be solved. The final step to reduce the computation time relies on the fact that ATHENA3D has been parallelized and adapted to the use of HPC resources. In this paper, the validation of the 3D FE model is discussed. A cross-validation of ATHENA 3D and CIVA is proposed for several inspection configurations. The performances in terms of calculation time are also presented in the cases of both local computer and computation cluster use.

1. Introduction

For obvious safety reasons, maintenance and In-service inspection is of great importance for nuclear power plants operators like EDF. Ultrasonic techniques are in particular very useful to detect and size potential flaws localized in depth in the metallic components. In these techniques, the detection is mostly based on amplitude criterion of the reflected signal from the defect. But, in some cases, due to unfavorable wave to microstructure interaction, the performances of the control can decrease significantly [1, 2]. Furthermore, the use of one NDT method for onsite nuclear inspection is conditioned by a qualification procedure which has to demonstrate the performances of the chosen technique. In this context, modeling codes can be of a great interest since they allow parametric studies [3] and avoid providing expensive mock-up. However, a good accuracy of the numerical results necessitates the taking into account of all physical phenomena involved in the wave propagation in polycrystalline media. In particular Finite Element (FE) codes have proved their efficiency to accurately reproduce complex phenomena such as beam deviation, division or attenuation [4-7]. During the past decade, EDF R&D developed in collaboration with INRIA the FE code ATHENA2D dedicated to the solving of the elastodynamics equations in all kind of elastic media [4, 8-10]. This code has already been widely used in order to simulate complex inspection configurations and is consequently relatively well documented. Nevertheless, up to now, only the 2D version of the code was available since an important amount of inspection configurations can be approximated by 2D models and 3D finite element (FE) models were generally considered to be prohibitively time consuming. But, in some cases, 3D simulations are required. However, the advances in computing processor speed and memory allow them to be more and more competitive and lead EDF R&D to develop the 3D version of the ATHENA code.

In this paper, the main characteristics of the code as well as the chosen numerical scheme are described. The performances in terms of calculation time are also presented in the cases of both local computer and calculation cluster use. We then present some cross-code validation configurations between the FE code ATHENA and the semi-analytical code CIVA. **2. Mathematical model**

ATHENA is a finite element code that simulates wave propagation in all kind of elastic media by solving the elastodynamic equations expressed in a mixed formulation combining stress and velocity terms (σ and v respectively).

Equations (1) and (2) are solved on a 2D or 3D calculation domain Ω whose boundary is ∂Ω and respects condition (3). Furthermore, a defect Γ can be included inside Ω as long as the free surface boundary condition (4) is satisfied. The particularity of the code relies on the fact that the interaction between the beam and the defect is simulated using the fictitious domains method. In this case, the defect is discretized with a triangular mesh (in the 3D case, with segments in the 2D case) independent of the 3D Cartesian mesh of Ω and allows arbitrary shape and orientation of the defect [11, 12].

(1) $\rho\, \partial v/\partial t - \mathrm{div}\sigma = f$,
(2) $A\, \partial\sigma/\partial t - \varepsilon(v) = 0$ on Ω,

(3) $\sigma.n = 0$ on $\partial\Omega$
(4) $\sigma.n = 0$ on Γ

where, v is the velocity in Ω, σ is the stress tensor, $\varepsilon_{i,j}(v) = ½(v_{i,j} + v_{j,i})$, $A = C^{-1,}$ with C the elasticity tensor and ρ is the density.

In addition, ATHENA gives the possibility to use perfectly matched absorbing layers (PML) to define the boundaries of the calculation domain in order to avoid parasite reflections on the artificial edges of the calculation zone and allow simulation in infinite domains.

3. Numerical Scheme

The discretization of (1) and (2) using of an appropriate variational formulation leads to the following numerical scheme:

(5) $$M_v \frac{V^{n+1} - V^n}{dt} + B\, \sigma^{n+1/2} = F^{n+1/2}$$

(6) $$M_s. \frac{\sigma^{n+1/2} - \sigma^{n-1/2}}{dt} + B^T.V^n - D^T.\Lambda^n = 0.$$

(7) $$D.\, M_s^{-1}.D^T.\Lambda^{n} = D.\, M_s^{-1}\, B^T\, .V^n$$

The notation V^n is used to represent V at time step n*dt. The above-written equations introduce a new quantity Λ, which is a Lagrange multiplier used to ensure the boundary condition (4). Mass matrices M_v and M_s depend respectively on the quantities ρ and A seen above and are block diagonal making the numerical scheme quasi-explicit. Matrix B is a discrete divergence operator and D is "trace operator" of the defect mesh on the domain mesh. Concerning the time discretization, we can notice that σ and V are not computed at the same time, (5) and (6) belongs to the class of "leap-frog" schemes.

4. Space discretization

The calculation domain Ω where equations (5) to (7) are solved is limited to rectangular surfaces in the 2D case and to rectangular parallelepipeds in the 3D case so that a regular Cartesian mesh is used for the discretization. The computation of V is performed in the middle of the cells (plane-rectangular or cubes) while σ is calculated at the nodes. However we do not compute one value of σ at each node. The choice of the finite elements leads to discontinuities at each cell for the different components of tensor σ. Indeed if we consider the 3D case: on one hand, σ_{xy} is continuous in both directions x and y, and discontinuous in direction z. On the other hand σ_{zz} is only continuous in direction z. So we have to compute 18 stress degrees of freedom on each node. We only have to solve a small system of 18 equations on each node, which is less time consuming than a global implicit system which would be a 21*Nx*Ny*Nz matrix.

As the mesh is regular, the stability of the explicit scheme can be easily ensured with $h = V_p * dt$, with V_p being the longitudinal velocity, dt the time step and h the mesh size.

The additional unknowns Λ, introduced by the fictitious domains method on every nodes of the defect mesh, can be obtained by solving a linear system at each time step. The order of the corresponding matrix is the number N of nodes of the defect mesh. Furthermore this matrix is constant and factorized once before the transient calculus.

5. Parallelization

Even if ATHENA does not deal with a global linear system of equations, the computation time of 3D configurations can become prohibitive in a local computer. In order to increase the competiveness of the code, it has been parallelized using the send/receive mechanism of the well-documented MPI (Message Passing Interface) library. In the parallelized version of ATHENA, the domain Ω is divided along Z axis in np chunks of size Nz/np, where np is the number of processors used. Two processors, in charge of two subdomains, can then communicate, sending and receiving data belonging to their common boundary (Figure 1). Furthermore, one can notice that there is no limitation for the defect due to parallelization since the defect can belong to more than one subdomain.

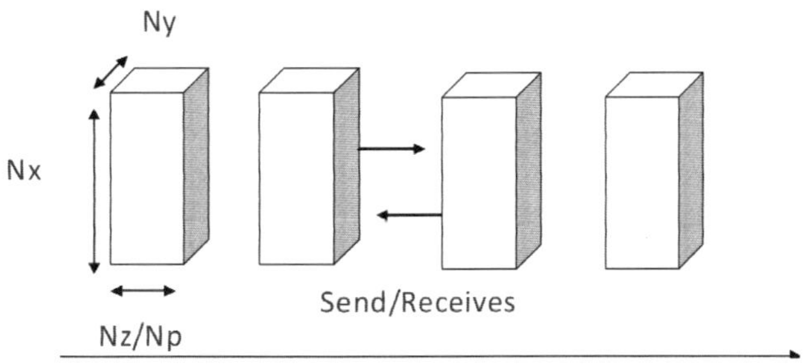

Figure 1. Communication scheme of two processors in Athena3d parallelization.

6. Computation time performances

According to the sizes of problems, ATHENA3D can be run on different kinds of computers. Small size problems can run on local machines. Indeed the development of multi-cores processors gives the possibility to take advantage of parallelization with 8 or 12 processors on local computers. However, for bigger size problems, the use of a calculation cluster becomes necessary. Table 1 summarized the computing times obtained on different machines for three different test cases with regard to their main numerical characteristics (number of cells, number of time steps and mesh size). However, it only gives an assessment of computing resources needed, and are not "best effort" performance times obtained by optimization.

The test cases are the followings:

- Case 1 corresponds to the inspection of a 10mm high notch located at 20 mm depth in stainless steel with a 0.2 mm mesh size
- Case 2 corresponds to the inspection of a 0.6 mm diameter SDH located at 30 mm depth in stainless steel and a mesh size of 0.08mm.
- Case 3 is similar to case 2 but with a 0.04mm mesh size

The local computer used for the time performances evaluation includes two Intel processors (Nehalem type, frequency: 2.3 GHz), with 12 Gbytes of memory. Eight cores are then available on this computer.

The cluster used in this study contained about 2000 computing nodes. Each node includes two Intel "Westmere" processors (frequency: 2.8 GHz), providing 12 cores and exhibits 24 Gbytes of memory. On such a node parallel computing can be run with 12 processes on 12 cores.

Table 1. Different calculation times of ATHENA3D

	Number of cells	Number of time step	Mesh size (mm)	Time calculation Local computer	Time calculation cluster
Case1	4 10^6	735	0.2	20 minutes	-
Case2	75 10^6	3215	0.08	~18 H	9 Hours
Case3	200 10^6	5700	0.04	-	30 H

The mesh size of case 1 is 0.2 mm which is a common value for 2 MHz longitudinal wave. As a consequence, the number of cells is limited enough to allow a local calculation of the inspection configuration. In this study, Case 1 was calculated using the 8 cores available on the local machine.

For Case 2, the use of the calculation cluster decreases the computation time by a factor of two. However, the calculation was launched on only one node of the cluster, so the 9 hours of computing time were obtained with a limited 12 cores parallelization. It means that the computation could have been run on a bigger amount of cores implying a decrease of the computing time.

Case 3 needed a big amount of memory such as the calculation could not be run on a local machine. Furthermore, for the cluster calculation, the domain Ω had to be divided into 24 subdomains, each one needing an entire computing node, and, as a result, used only one of the 12 cores. This is a "memory driven" case. Finally a 200 million cells calculation has been calculated in 30h.

Nevertheless, it has to noted that the computation time performances were evaluated for a configuration with only one Ascan, i.e. one position of the UT probe. When an entire Bscan must be simulated, the use of the calculation cluster allows a more than substantial gain of time since each Ascan can be launched simultaneously while they have to be calculated sequentially in a local machine.

7. Athena pre/post processing

The utilization of ATHENA is simplified by the integrated Graphical User Interface (GUI): MILENAQt (Figure 2). It helps the user to create the simulation configuration, i.e. the calculation domain, defect shape and size, probe characteristics, etc., through the existence of various menus and submenus guiding the user step by step. Furthermore, it includes a preprocessing stage consisting in an automatic meshing of the defect and calculation domain and the computation of the source field using the probe characteristics defined by the user. Finally, MILENAQt integrates the post-processing and the visualization of the results (Figure 3).

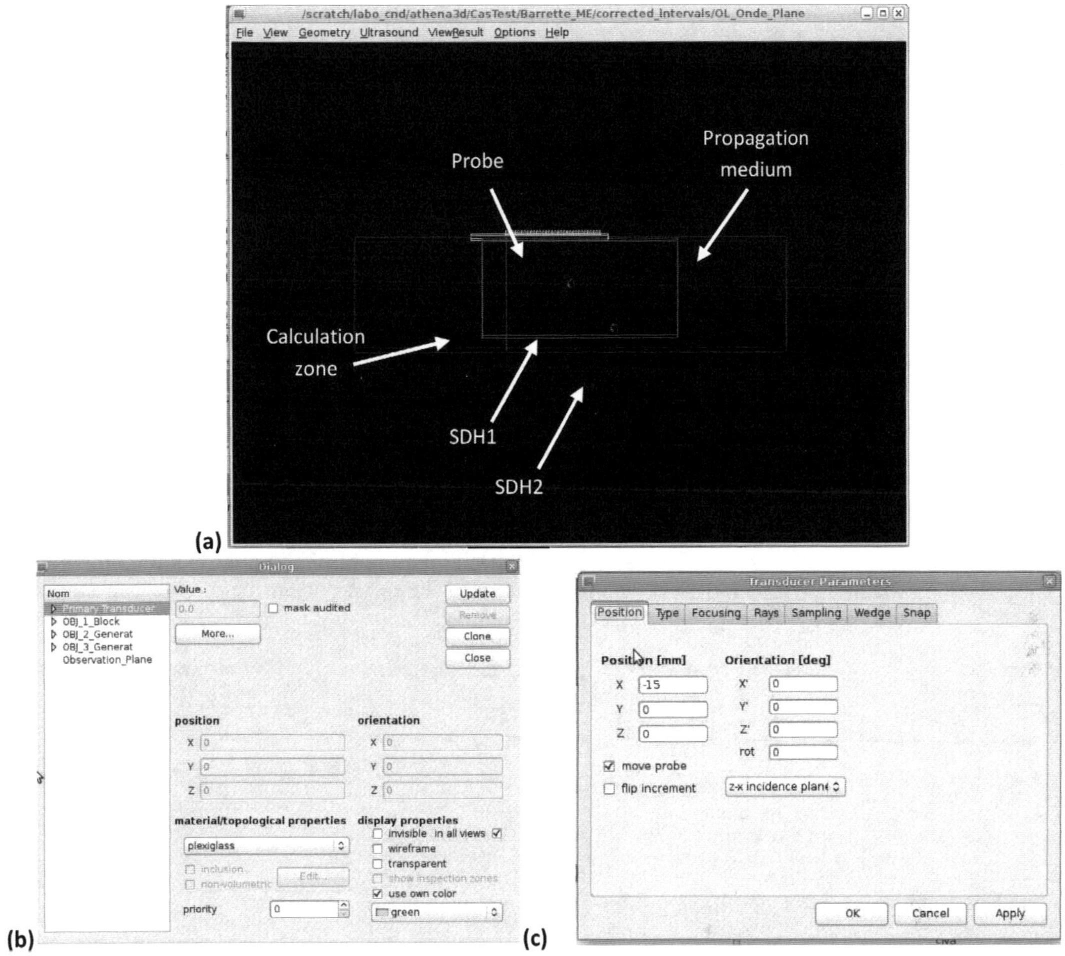

Figure 2. Illustration of the GUI MilenaQt. (a) shows the visualization of the inspection configuration: two SDH and a linear phased array probe. (b) shows the dialog box dedicated to the definition of the inspected object (propagation media and defects). (c) shows the dialog box dedicated to definition of the transducer parameters.

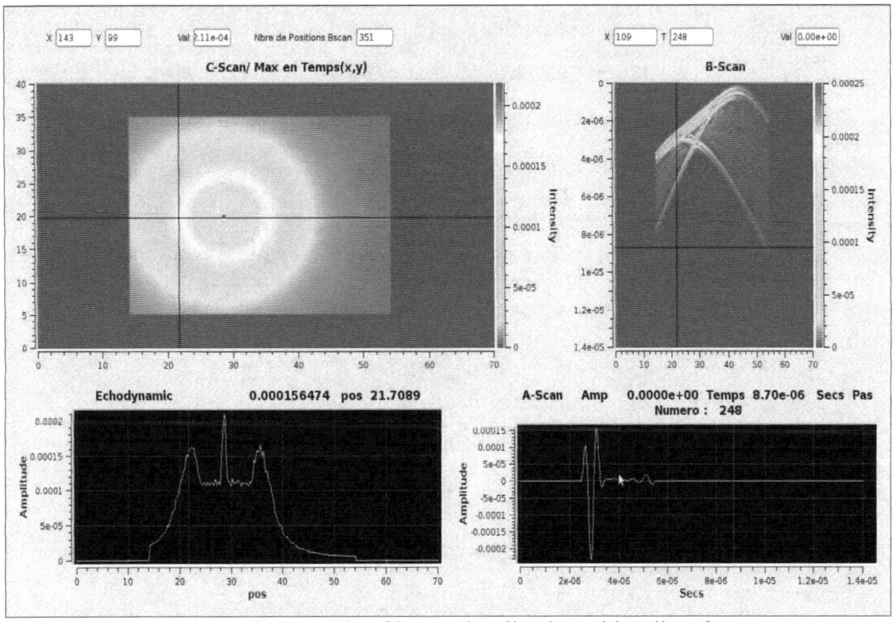

Figure 3. Example of beam visualization with MilenaQt.

8. Test configurations in a homogeneous and isotropic medium

In this study, three validation cases were scrutinized. In each case, the propagation medium made of austenitic stainless steel is isotropic and homogeneous (V_L= 5 700 m/s, V_S= 3130 m/s and its density is d=7.9).

The first configuration involved the detection of two side drilled holes (SDH) located at 10 and 20 mm depth with a 1.5 mm diameter. Four different single element 2MHz probes were simulated. Their characteristics are summarized in Table 2. An example of configuration is given Figure 4.

The amplitudes of the SDH echoes obtained with ATHENA3D were compared with the ones derived from CIVA10 simulations performed with the same inspection description. The results are summarized in Table 3. The amplitude of the 10 mm depth SDH was used as the reference.

Table 3 shows a good agreement between the two codes. Indeed, except for the fL55 law for whom the discrepancy is 1.5 dB, the differences between ATHENA3D and CIVA 10 are inferior to 0.5 dB. Furthermore, the focalization effects are correctly taking into account such as the amplitude of the 20 mm depth increases significantly when using the appropriate focal law (fL0 case). In addition, Figure 5 shows the fields obtained for the two focalized probes for both ATHENA3D and CIVA 10. They exhibit very similar properties and in particular, they provide the same focusing depths.

Table 2. Characteristics of the probes used for the first validation case

Probe	Element shape	Propagation angle in austenitic steel (°)	Inspection type
L0	Cylindrical R= 6mm	0	Contact (wedge 5mm high) Plane L wave
L55	Cylindrical R= 6mm	55	Contact (wedge 15mm high) Plane L wave
fL0	Spherical Curvature radius = 150 mm R= 20 mm	0	Immersion (water height =50 mm) Focused L wave (focalization depth = 22 mm)
fL55	Spherical Curvature radius = 150 mm R= 20 mm	55	Immersion (water height =50 mm) Focused L wave (focalization depth = 12 mm)

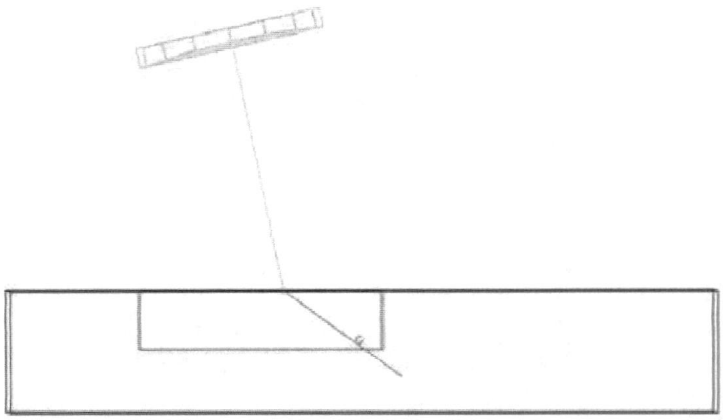

Figure 4. Inspection configuration of a 1.5 mm diameter of SDH located at 10 mm depth with a focused mono-element probe at the propagation angle of 55°.

Table 3. Amplitudes of the 20 mm depth SDH echo normalized by the 10 mm depth SDH obtained with ATHENA3D and CIVA 10

	L0 (dB)	L55 (dB)	fL0 (dB)	fL55 (dB)
ATHENA3D	-1.5	-6.0	+15.0	-4.0
CIVA 10	-1.5	-6.0	+15.5	-2.5

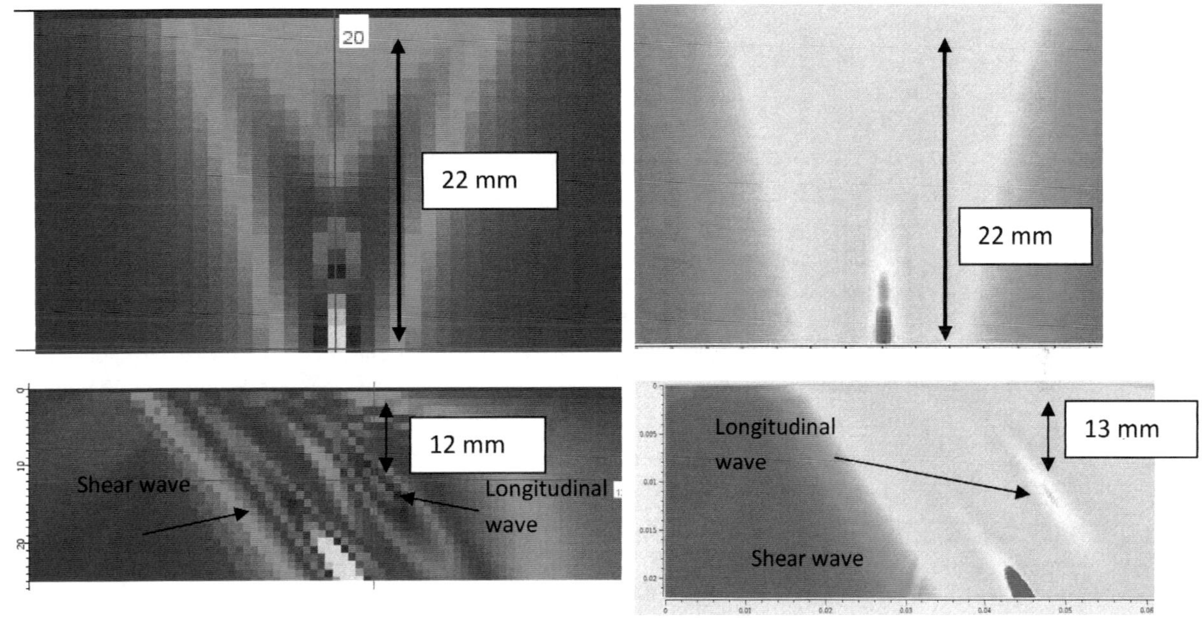

Figure 5. Sound field for the fL0 and fL55 calculated using CIVA 10 (left) and ATHENA3D (right).

The second configuration corresponds to the detection of a 10 mm height notch with a TRL55 probe (separate transmitter/receiver probe, longitudinal wave at 55° of incidence and 2MHz central frequency). The inspection configuration is given Figure 6. The Bscan obtained using ATHENA3D (25 scans with 1 mm step) can be observed on Figure 7. The main expected phenomena are clearly visible: the diffraction, the corner and the LLT mode conversion echoes are present. As for the previous configuration, the amplitudes of these echoes were compared to those computed using CIVA10. The results are summarized in Table 4. Again, a good agreement between CIVA and ATHENA was observed since the discrepancy between the two models is inferior to 2 dB.

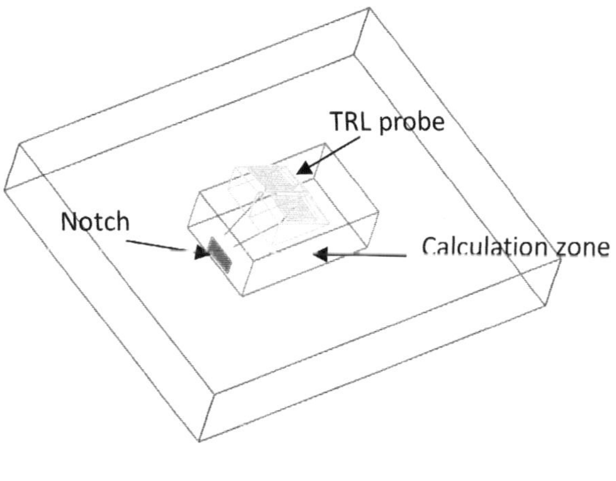

Figure 6. Inspection of a 20 mm high notch with a TRL probe (incidence angle of 55°)

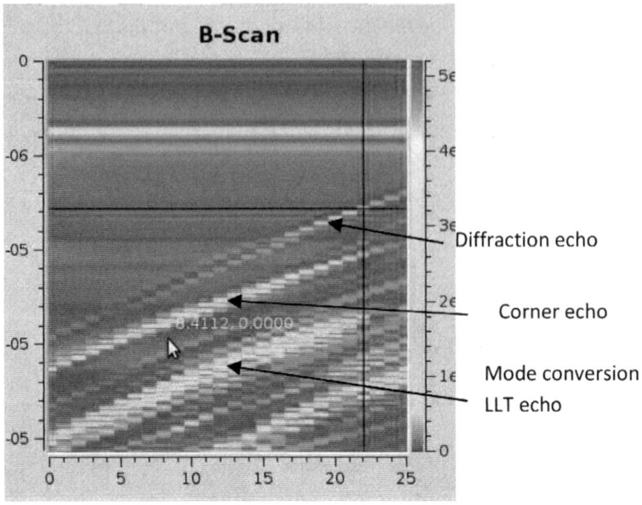

Figure 7. Bscan corresponding to the inspection of a notch with a TRL probe. The diffraction, corner and mode conversion echoes are clearly visible.

Table 4. Amplitude of the main echoes coming from the notch inspection. The reference is a 1.5 mm diameter SDH located at a 30 mm depth

	Corner (dB)	Diffraction (dB)	LLT (dB)
ATHENA3D	3.0	-6.0	13.5
CIVA 10	2.5	-8.0	11.5

The third configuration concerns the use of a linear phased array (PA) probe for the detection of 1.5 mm diameter SDHs. A 32 elements linear probe is simulated. Its properties are described in Table 5. The inspection configuration is illustrated Figure 8. Three focal laws were tested: L0 plane wave, L0 with a 10 mm focusing depth (0F10) and L0 with a 20 mm focusing depth (0F20).

Table 5. Properties of the simulated linear phased array

Central frequency	Element size	Array pitch	Number of element
2 MHz	0.3x15 mm^2	0.5 mm	32

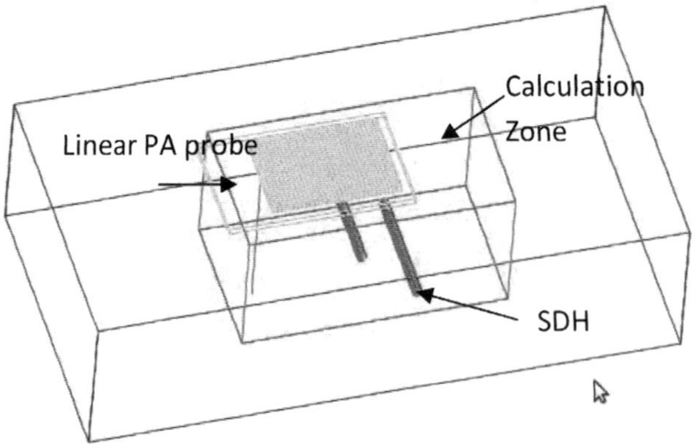

Figure 8. Inspection configuration of 2 SDH using a linear PA probe

The sound fields simulated for the 0F10 and 0F20 laws using CIVA 10 and ATHENA are given Figure 9 and Figure 10. In both cases, the focal laws are clearly taken into account. Furthermore, the two codes give similar results in terms of focal depth.

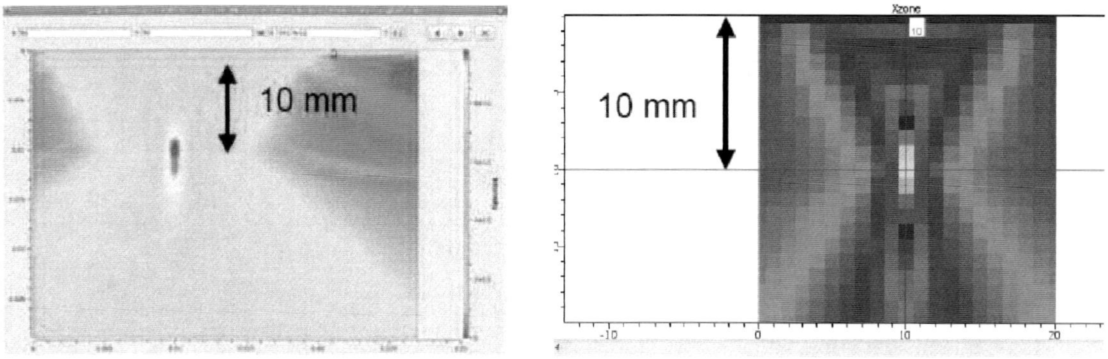

Figure 9. Sound field calculated using ATHENA3D (left) and CIVA 10 (right) for the 0F10 focal law

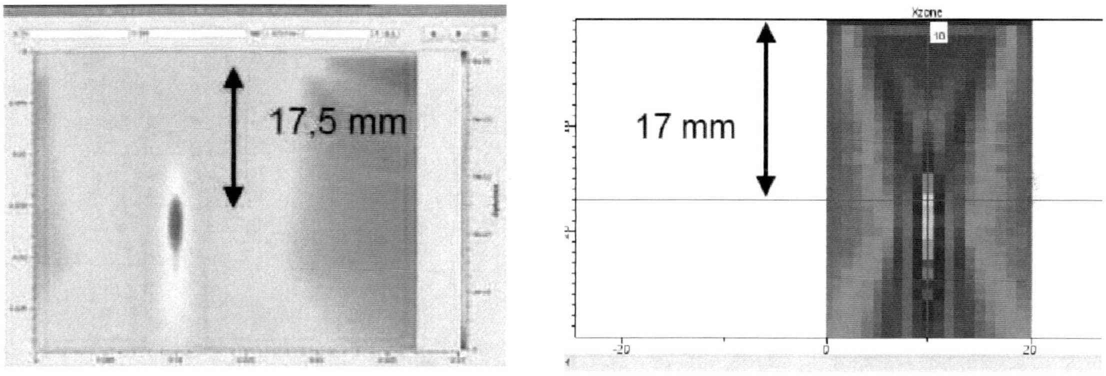

Figure 10. Sound field calculated using ATHENA3D (left) and CIVA 10 (right) for the 0F20 focal law

Concerning the detection of the SDHs, the amplitude of the 20 mm depth SDH obtained for the 3 focal laws and the 2 codes are summarized in Table 6. The discrepancy between the two remains inferior to 2 dB which is a satisfactory result in terms of cross-code validation.

Table 6. Amplitude (dB) of the SDH located at 20 mm depth. The reference is the amplitude of the 10 mm depth SDH measured with the focal law 0F10.

	Plane Wave	0F10	0F20
ATHENA 3D	-10.0	-7.5	-1.5
CIVA 10	-11.5	-8.5	-3.5

9. Anisotropic medium

A preliminary test has been realized to verify the capacity of the ATHENA3D code to take into account beam deviation when it propagates in an anisotropic (but homogeneous) medium. In this configuration, the simulated medium exhibits a density of 7 and transversally isotropic elastic properties (Table 7) and the probe provides a longitudinal wave with no incidence angle (L0). We want to evaluate the influence of the orientation of the principal symmetry axis on the beam propagation. To do so, three different cases are tested. In the first one, the symmetry axis is parallel to the propagation axis (z). In the second case, the symmetry axis undergoes a 15° rotation around the y-axis. In the third case, the symmetry axis is rotated by 15° around the x-axis (Figure 11).

For the three cases, the simulated sound field on an observation plane located at a 30 mm depth from the entry surface is given in Figure 12. The results are in very good agreement with the theory of ultrasound propagation in anisotropic media. Indeed, in the first case, no deviation is observed while in the second and third cases, an 8.5 mm deviation along the x axis and y axis (respectively) appears.

Table 7. Elastic properties of the transversally isotropic austenitic stainless steel used for the simulation

C_{11}	C_{22}	C_{33}	C_{44}	C_{55}	C_{66}	C_{12}	C_{13}	C_{23}
247	247	218	105	105	80	105	148	148

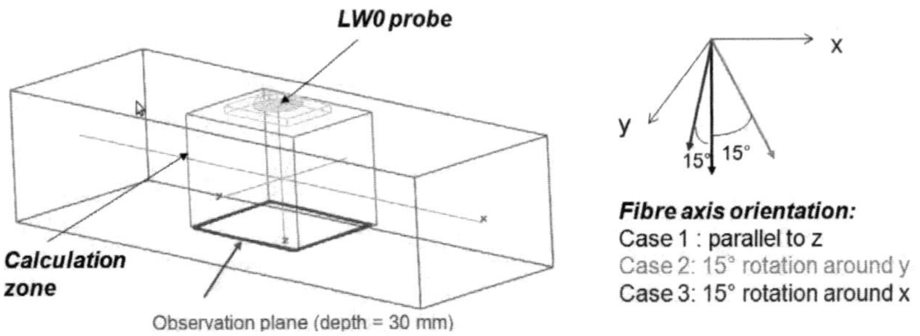

Figure 11. Computation configuration for the 3 orientations of the symmetry axis (Fiber orientation) of the medium

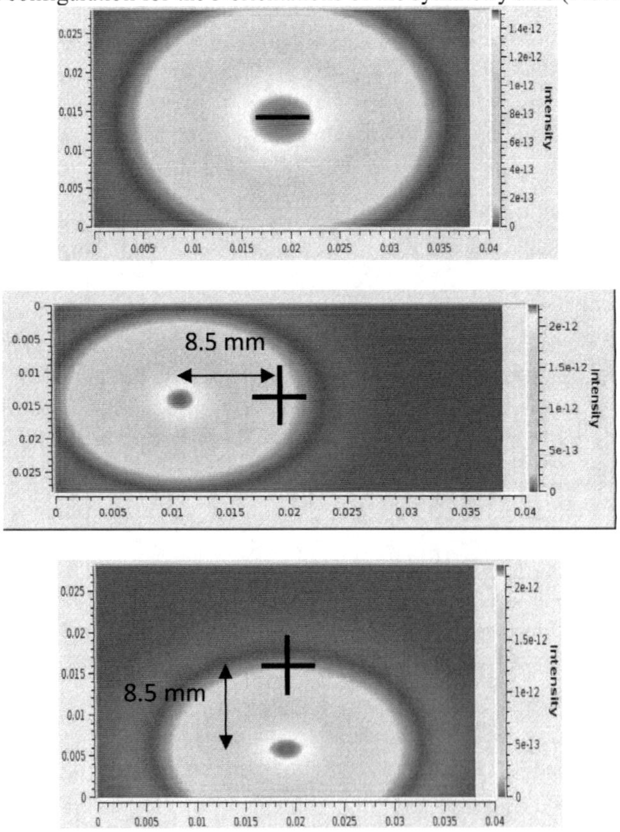

Figure 12. sound field calculated at the observation plane using ATHENA 3D when the symmetry axis is a) parallel to z, b) rotated around the y axis and c) rotated around the x axis

10. Conclusion

In the present paper, the new version of the ATHENA code is presented. It allows 3D finite element calculation and is optimized for parallelized computation. Consequently, in numerous configurations, the use of such a code is no more prohibitively time consuming when launching the computation on a calculation cluster.

Furthermore, the first validations cases detailed in this work showed that the ATHENA code was able to reproduce both wave propagation and wave to defect interaction in isotropic and homogeneous media. In addition, various kinds of ultrasonic probes (single and multi-element probes, immersion, contact, plane wave, focalized wave, etc.) can be handled by ATHENA through the use of its GUI MILENAQt.

However, this validation work must be continued. Indeed, simulation of Distance-Amplitude Curve (DAC) covering a wide range of depth is needed in order to quantitatively define the validation domain of the code. Furthermore, in this study, the comparison between the sound fields obtained from CIVA 10 and ATHENA 3D was only based on the determination of the focal depth. Consequently, a more exhaustive comparison of the sound fields providing by each code should be of great interest.

Finally, even if a preliminary result obtained in an anisotropic homogeneous medium has been shown in this paper, it is necessary to perform simulation on heterogeneous structures such as austenitic welds.

References

[1] Chassignole B, Villard D, Dubuget M, Baboux J-C and El Guerjouma R 2000 Characterization of austenitic stainless steel welds for ultrasonic NDT *Review of Progress in QNDE* **19** AIP Conference Proceedings, **509**, 1325-32

[2] Feuilly N, Dupond O, Chassignole B, Moysan J and Comeloup G 2009 Relation between ultrasonic backscattering and microstructure for polycrystalline materials *Review of Progress in QNDE* **28** AIP Conference Proceedings, **1096**, 1216-23

[3] Rupin F, Blatman G, Lacaze S and Fouquet T 2013 Sensitivity analysis of an ultrasonic inspection of a weld structure simulated with the finite element code ATHENA 2D *Review of Progress in QNDE* AIP Conference Proceedings, **1511**, 1089-96

[4] Chassignole B, Duwig V, Ploix M-A, Guy P and El Guerjouma R 2009 Modelling the attenuation in the ATHENA finite elements code for the ultrasonic testing of austenitic stainless steel welds *Ultrasonics* **49** 653-8

[5] Chassignole B, El Guerjouma R, Ploix M-A, Fouquet T 2010 Ultrasonic and structural characterization of anisotropic austenitic stainless steel welds: Towards a higher reliability in ultrasonic non-destructive testing *NDT&E Int.* **43** 273-82

[6] Chassignole B, Dupond O, Doudet L, Duwig V and Etchegaray N 2009 Ultrasonic examination of an austenitic weld: illustration of the disturbances of the ultrasonic beam *Review of Progress in QNDE* **28**, AIP Conference Proceedings, **1096**, 1886-93

[7] Hannemann R, Marklein R, Langenberg K J, Schurig C, Köhler B and Walte F 2000 Ultrasonic wave propagation in real-life austenitic V-Butt welds: numerical modeling and validation*Review of Progress in QNDE* **19** AIP Conference Proceedings **509**, 145-52

[8] Apfel A, Moysan J, Corneloup G, Fouquet T and Chassignole B 2005 Coupling an ultrasonic propagation code with a model of the heterogeneity of multipass welds to simulate ultrasonic testing *Ultrasonics* **43** 447-56

[9] Chassignole B, Diaz J, Duwig V, Fouquet T and Schumm A 2006 Structural noise in modelisation *in ECNDT 2006 proceedings* Fr.1.4.4, 1-13

[10] Shahjahana S, Rupin F, Fouquet T, Aubryc A and Derode A 2012 Structural noise and coherent backscattering modelled with the ATHENA2D finite element code *in Acoustics 2012*, Nantes, 2639-44

[11] Bécache E, Joly P and Tsogka C 2001 Fictitious domains, mixed finite elements and perfectly matched layers for 2D elastic wave propagation *J. Comp. Acoustics* **9** 1175-201

[12] Bécache E, Joly P and Tsogka C 2000 An analysis of new mixed finite elements for the approximation of wave propagation problems *SIAM J. Num. Anal.* **37**, 1053-84

Acknowledgement

This work was realized in the framework of the MOSAICS project (Modeling of an austenitic stainless steel weld inspected by ultrasonic techniques) which is supported by the French National Agency of Research.

Journal of Physics: Conference Series **498** (2014) 012010 doi:10.1088/1742-6596/498/1/012010

Edge Diffraction Coefficients around Critical Rays

L Fradkin[1], M Harmer[1] and M Darmon[2]

[1] Sound Mathematics Ltd., 11 Mulberry Close, Cambridge CB4 2AS, UK
[2] CEA, LIST, Department of Imaging and Simulation for NDT, F-91191 Gif-sur-Yvette, France

E-mail: l.fradkin@soundmathematics.com

Abstract. The classical GTD (Geometrical Theory of Diffraction) gives a recipe, based on high-frequency asymptotics, for calculating edge diffraction coefficients in the geometrical regions where only diffracted waves propagate. The Uniform GTD extends this recipe to transition zones between irradiated and silent regions, known as penumbra. For many industrial materials, e.g. steels, and frequencies utlized in industrial ultrasonic transducers, that is, around 5 MHz, asymptotics suggested for description of geometrical regions supporting the head waves or transition regions surrounding their boundaries, known as critical rays, prove unsatisfactory. We present a numerical extension of GTD, which is based on a regularized, variable step Simpson's method for evaluating the edge diffraction coefficients in the regions of interference between head waves, diffracted waves and/or reflected waves. In mathematical terms, these are the regions of coalescence of three critical points — a branch point, stationary point and/or pole, respectively. We show that away from the shadow boundaries, near the critical rays the GTD still produces correct values of the edge diffraction coefficients.

1. Introduction

The geometrical theory of diffraction (GTD) is an extension of the standard ray theory. According to GTD, in the high frequency approximation, i.e. under the assumption that the distance from the observation point to the edge diffrcation point as well as both the curvature of the wave front and crack edge at this point are larger than the wavelength, a wave incident on a sharp edge produces a cone of diffracted rays. Diffraction coefficients describe the directivity patterns of the resulting diffracted fields.

GTD is used widely in a variety of applications and in particular, in ultrasonic NDE (Non-Destructive Evaluation). Formulae for evaluating elastodynamic diffraction coefficients for a semi-infinite planar crack in an isotropic homogeneous solid have been presented e.g. in [1] and [2]. A code based

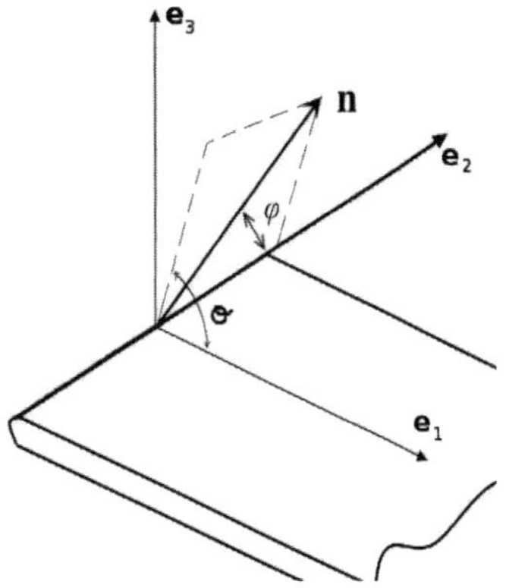

Figure 1. The coordinate systems associated with the crack.

on these formulae was incorporated into CIVA, an NDE simulation software developed by CEA [3]. A new simpler version of these formulae has been presented recently in Gautesen [4] and also incorporated into CIVA [5].

The aim of this paper is to describe a method for evaluating the diffraction coefficients of the half-plane crack in an isotropic homogeneous solid — in the regions of interference between head waves, diffracted waves and/or reflected waves, where the standard GTD fails. In mathematical terms, the regions under consideration are regions of coalescence of three critical points — a pole, stationary point and branch point, respectively. The evaluation is carried out numerically, using a regularised, variable step version of Simpson's method.

2. Statement of the canonical edge diffraction problem

Let us assume that the three-dimensional space is filled with an isotropic homogeneous solid supporting two speeds, c_L, the speed of longitudinal waves and c_T, the speed of transverse waves. In the presence of a scatterer, $\mathbf{u}$, the total displacement field in this solid can be decomposed into the sum of the incident field $\mathbf{u}^i$ and scattered field, $\mathbf{u}^{sc}$. In the high frequency approximation, the field scattered by a scatterer with an edge is the sum of the reflected field $\mathbf{u}^r$ and the diffracted field $\mathbf{u}^d$, which describes the effect of the edge, so we can write $\mathbf{u} = \mathbf{u}^i + \mathbf{u}^{sc} = \mathbf{u}^i + \mathbf{u}^r + \mathbf{u}^d$. Each of the components of the geometrical elastodynamic field $\mathbf{u}^i + \mathbf{u}^r$ vanishes in the respective shadow (silent) zone. A transition zone in between an irradiated region and the corresponding silent zone is called a penumbra.

Let us now assume that the scatterer is a semi-infinite planar crack. Let us introduce the Cartesian coordinate system such that it lies in the (x_1, x_2)-plane. Let the crack edge coincide with the x_2-axis and let the crack contain the positive x_1-axis (see Fig. 1). Let it be irradiated by a time-harmonic wave represented by $\mathbf{U}(\mathbf{x}, t) = \mathbf{u}(\mathbf{x}) \exp(i\omega t)$, where i is the imaginary unity; $\mathbf{x} = (x_1, x_2, x_3)$ is an observation point; t the time; and ω is the circular frequency. Below only time-harmonic waves are considered and the factor

$\exp(i\omega t)$ is implied but omitted everywhere.

The following conventions are followed above and everywhere below: The bold font is used to denote vector quantities. Whenever appropriate, these symbols are supplied with superscripts to describe the nature of the wave. In particular, the descriptors i and d introduced above (and sometimes put in curly brackets below) denote, respectively, the incident and diffracted waves. The descriptors α, $\beta = L, TH, TV$ or R introduced below refer to longitudinal, transverse horizontal, transverse vertical or Rayleigh wave, respectively, with α normally used to denote the incident wave mode and β, the scattered one. Subscripts are used to refer to the corresponding vector components or, when dealing with scalars, to the wave mode.

To continue, the equation of motion for this problem can be written as

$$\nabla \cdot \sigma(\mathbf{x}) = \rho \frac{\partial^2 \mathbf{u}}{\partial t^2}(\mathbf{x}), \tag{1}$$

where $\nabla = (\partial/\partial x_1, \partial/\partial x_2, \partial/\partial x_3)$ is the nabla operator with ∂ - the symbol of partial differentiation; $\sigma(\mathbf{x})$ is the stress tensor; and ρ is the density of the medium. Using Hooke's law, the Cartesian components of the stress tensor are

$$\sigma_{ij}(\mathbf{x}) = \lambda \delta_{ij} \frac{\partial u_k}{\partial x_k}(\mathbf{x}) + \mu \left(\frac{\partial u_i}{\partial x_j}(\mathbf{x}) + \frac{\partial u_j}{\partial x_i}(\mathbf{x}) \right), \quad i, j = 1, 2, 3, \tag{2}$$

where λ and μ are the Lamé parameters and δ_{ij} is the Kronecker delta.

A realistic boundary condition [2] prescribes traction-free crack faces,

$$\mathbf{t}(\mathbf{x})|_{crack} = \mathbf{0}, \tag{3}$$

where $\mathbf{t}(\mathbf{x}) = \boldsymbol{\nu} \cdot \sigma(\mathbf{x})$ and $\boldsymbol{\nu}$ is the inner normal to the upper face pointing into the solid. It is further assumed that the scattered waves satisfy the standard radiation condition at infinity demanding that they be outgoing.

3. Solution of the canonical edge diffraction problem

Solving the above problem shows that the isotropic medium supports two types of plane waves $\mathbf{u} = A\mathbf{d}\exp(ik\mathbf{p}\cdot\mathbf{x})$, L (longitudinal), also known as P (compressional) waves and T (transverse), also known as S (shear) waves. The polarization vector (unit displacement vector) $\mathbf{d}^L$ is parallel to propagation vector $\mathbf{p}$. The T wave is decomposed into a TV vertical component and a TH horizontal component with respect to a plane. Choosing this plane to be a crack plane and using the decomposition in [4], the corresponding polarization vectors are

$$\mathbf{d}^{TV} = \frac{\mathbf{p} \times \mathbf{e}_2}{|\mathbf{p} \times \mathbf{e}_2|}, \quad \mathbf{d}^{TH} = \mathbf{p} \times \mathbf{d}^{TV} \tag{4}$$

It is shown in [4] that the single Fourier transform of the scattered field $\hat{\mathbf{u}}^{sc}(\xi_1, x_2, x_3)$ can be represented as

$$\hat{\mathbf{u}}^{sc}(\xi_1, x_2, x_3) = \frac{-i\kappa_\beta^2}{\kappa_T^2} \sum_\beta \frac{1}{\gamma_\beta(\xi_1)} \times [\mathbf{v}^{\beta\{d\}}(\hat{\boldsymbol{\xi}}) \cdot \mathbf{v}^{\alpha\{i\}}(\boldsymbol{\xi}^{\alpha\{i\}})] \times \mathbf{d}^{\beta\{d\}}(\hat{\boldsymbol{\xi}}) \times e^{ik_L[\xi_2 x_2 + |x_3|\gamma_\beta(\xi_1)]}, \tag{5}$$

where the single Fourier transform of $f(x_1)$ is

$$\hat{f}(\xi_1) = k_L \int_{-\infty}^{\infty} f(x_1)\, e^{ik_L\xi_1 x_1} dx_1; \tag{6}$$

$\beta = L, TV, TH$; $\kappa_L = 1$; $\kappa_T = c_L/c_T$; $\boldsymbol{\xi} = (\xi_1, \xi_2, \xi_3) = \kappa\mathbf{p}$ is the dimensionless wave vector; $\hat{\boldsymbol{\xi}} = (-\xi_1, \xi_2, s\gamma_\beta(\xi_1))$ is the auxulliary wave vector; $s = sign(x_3)$, with $\gamma_\beta(\xi_1) = \sqrt{\kappa_\beta^2 - \xi_1^2 - \xi_2^2}$; the expressions for vectors $\mathbf{d}$ and $\mathbf{v}$ can be found in [4] and are not reproduced here. We just emphasize that they depend on the functions $\gamma_\beta^+(\xi_1) = \sqrt{\xi_1 + \sqrt{\kappa_\beta^2 - \xi_2^2}}$. The scattered field can be evaluated using the Inverse Fourier Transform

$$\mathbf{u}^{sc}(\mathbf{x}) = \frac{k_L}{2\pi} \int_{-\infty}^{\infty} \hat{\mathbf{u}}^{sc}(\xi_1, x_2, x_3)\, e^{-ik_L\xi_1 x_1} d\xi_1. \tag{7}$$

In the high-frequency approximation the above integrand is a product of a fast-oscillating exponential factor and often a slowly varying amplitude. Such integrals can often be evaluated asymptotically and when the phase stationary point is isolated each diffracted field can be expressed as

$$\mathbf{u}^{\beta\{d\}}(\mathbf{x}) \simeq \frac{\mathrm{e}^{ik_L\boldsymbol{\xi}^{\alpha\{i\}}\cdot\mathbf{x}^{\beta\{d\}}}}{\sqrt{\kappa_\beta r^\beta}} S_\beta^\alpha\left(\boldsymbol{\xi}^{\beta\{d\}}, \boldsymbol{\xi}^{\alpha\{i\}}\right) \mathbf{d}^{\beta\{d\}}\left(\boldsymbol{\xi}^{\beta\{d\}}\right) \mathrm{e}^{ik_L\kappa_\beta r^\beta}, \tag{8}$$

where $\exp(ik_L\, \boldsymbol{\xi}^{\alpha\{i\}} \cdot \mathbf{x}^{\beta\{d\}})$ is the incident wave, the diffraction point $\mathbf{x}^{\beta\{d\}} = (0, x_2^{\beta\{d\}}, 0)$, the diffracted wave vector $\boldsymbol{\xi}^{\beta\{d\}} = (\xi_1^{\beta\{d\}}, \xi_2^{\beta\{d\}}, s\gamma_\beta(\xi_1^{\beta\{d\}}))$; r^β is the distance from the diffraction point to the observation point $\mathbf{x}$, so that we have $\kappa_\beta^{-1} r^\beta \boldsymbol{\xi}^{\beta\{d\}} = \mathbf{x} - \mathbf{x}^{\beta\{d\}}$ and the edge diffraction coefficient $S_\beta^\alpha = D_\beta^\alpha$ is

$$D_\beta^\alpha(\boldsymbol{\xi}^{\beta\{d\}}, \boldsymbol{\xi}^{\alpha\{i\}}) = \frac{e^{i\frac{\pi}{4}}\kappa_\beta^3 \mathbf{v}^{\beta\{d\}}(\boldsymbol{\xi}^{\beta\{d\}}) \cdot \mathbf{v}^{\alpha\{i\}}(\boldsymbol{\xi}^{\alpha\{i\}})}{\sqrt{2\pi}\kappa_T^2 \gamma_\beta(0)(\xi_1^{\beta\{d\}} - \xi_1^{\alpha\{i\}})}. \tag{9}$$

If the stationary point is not isolated and is coalescing with the branch point or pole or both, then comparing (7) and (8) and projecting (8) onto the displacement vector $\mathbf{d}^\beta(\boldsymbol{\xi}^\beta)$ we can still represent the scattered filed in the form (8) but in this case the scattering coefficient S_β^α can be calculated using

$$S_\beta^\alpha(\boldsymbol{\xi}^{\beta\{d\}}, \boldsymbol{\xi}^{\alpha\{i\}}) = \frac{-i\kappa_\beta^2}{2\pi\kappa_T^2} \int_{-\infty}^{\infty} \frac{\sqrt{\kappa_\beta r^\beta}}{\gamma_\beta(\xi_1)\left(\xi_1 + \xi_1^{\alpha\{i\}}\right)} \mathrm{e}^{iaf(\xi_1)} \times$$
$$[\mathbf{v}^{\beta\{d\}}(\hat{\boldsymbol{\xi}}) \cdot \mathbf{v}^{\alpha\{i\}}(\boldsymbol{\xi}^{\alpha\{i\}})] \times [\mathbf{d}^{\beta\{d\}}\left(\hat{\boldsymbol{\xi}}\right) \cdot \mathbf{d}^{\beta\{d\}}(\boldsymbol{\xi}^{\beta\{d\}})]\, d\xi_1, \tag{10}$$

where the far field parameter $a = k_\beta r^\beta$ and the phase is given by

$$f(\xi_1) = \kappa_\beta^{-2}\left(-\xi_1\, \xi_1^{\beta\{d\}} + \gamma_\beta(\xi_1)\left|\xi_3^{\beta\{d\}}\right| - \gamma_\beta^2(0)\right). \tag{11}$$

When the reflected waves, head waves and Rayleigh waves are removed from the integral in (10) what remains is the diffracted fields or else the fields which are due to the interference between diffracted and other fields.

Below we use $\xi_{st} = -\xi_1^{\beta\{d\}}$ for the stationary point, $\xi_p = -\xi_1^{\alpha\{i\}}$ for the incident pole and $\xi_{br} = -\gamma_L(0) = -\sqrt{1-\xi_2^2}$ for the branch point. We denote the angles of the incoming wave by ϕ^i and θ^i, and the angles for the scattered wave by ϕ^d and θ^d, using the spherical coordinate system shown in Fig. 1.

4. A new numerical integration technique

The integral in (10) can be evaluated using Simpson's rule adjusted to accommodate the presence of the fast oscillating exponent, a pole and branch point. We describe each of the aspects of the adjusted Simpson's method below.

4.1. Integration interval

Since we are interested in head waves and their interaction with other waves we concentrate on the T scattered field only. The integration in (10) is carried over the entire real axis and a typical integrand is presented in Fig. 2. We note that the phase $f(\xi_1)$ is real inside the interval $I = [-\gamma_T(0), \gamma_T(0)]$ and this interval contains both the stationary point $\xi_{st} = -\xi_1^{\beta\{d\}}$ and branch point $\xi_{br} = -\sqrt{1-\xi_2^2}$. The integrands in Fig. 2 display two other features:

- The endpoints of the interval I are accumulation points of the phase, in the sense that the derivative of the phase (11) increases without bound as we approach the endpoints, see Fig. 2, and when these end points are included numerical integration becomes unstable. Another reason to avoid the endpoints is the presence of $\gamma_T(\xi_1)$ in the denominator of the integrand in (10). Therefore, when the diffraction angle is away from grazing we carry out numerical integration over the interval

$$I_\delta = [-\gamma_T(0) + \delta_-, \gamma_T(0) - \delta_+], \tag{12}$$

  where the offsets $\delta_\pm = \delta \cdot \min(|\gamma_T(0) \mp \xi_{st}|, |\gamma_T(0) \mp \xi_{br}|)$ and δ is a prescribed small number. The choice assures that the main critical points ξ_{st} and ξ_{br} lie inside the integration interval, the excluded regions are small. The real and imaginary parts of the integrand oscillate around zero fast, thus making the neglected contribution very small. When the incident pole lies close to the lower/upper bound $-\gamma_T/\gamma_T$, we redefine $\delta_\pm$ as $\delta_\pm = 0.001\delta \cdot (|\gamma_T(0) \mp \xi_p|)$. When the diffraction angles are near grazing the interval contains no phase stationary point and we choose $\delta_\pm$ to be $\delta_\pm = \delta \cdot |\gamma_T(0) \mp \xi_p|$. Using numerical experimentation we have established that in both cases the most robust choice for δ is 0.1.

  When the interval I contains neither stationary point nor pole we choose $\delta_\pm = \pm 10^{-9}$. When all three critical points coalesce, numerical experimentation suggests the choice $\delta_\pm = \pm 10^{-6}$.
- The integrands decay exponentially outside the interval I. We make allowance for this by following advice given in [6], that is, integrating

by parts and neglecting terms of order $O\left(a^{-2}\right)$ to obtain the end point contribution (the estimate of the semi-infinite tails):

$$\begin{aligned}\mathcal{I}_{\infty} &= \int_{-\infty}^{\infty} H(\xi)\,\mathrm{e}^{iaf(\xi)}\,dx = \left[\int_{-\infty}^{\xi_-} + \int_{\xi_-}^{\xi_+} + \int_{\xi_+}^{\infty}\right] \frac{H(\xi)}{iaf'(\xi)}\, d\left(\mathrm{e}^{iaf(\xi)}\right) \\ &\approx \int_{\xi_-}^{\xi_+} H(\xi)\,\mathrm{e}^{iaf(\xi)}\,d\xi - \frac{1}{iaf'(\xi)} H(\xi)\,\mathrm{e}^{iaf(\xi)}\Big|_{\xi_-}^{\xi_+}. \end{aligned} \tag{13}$$

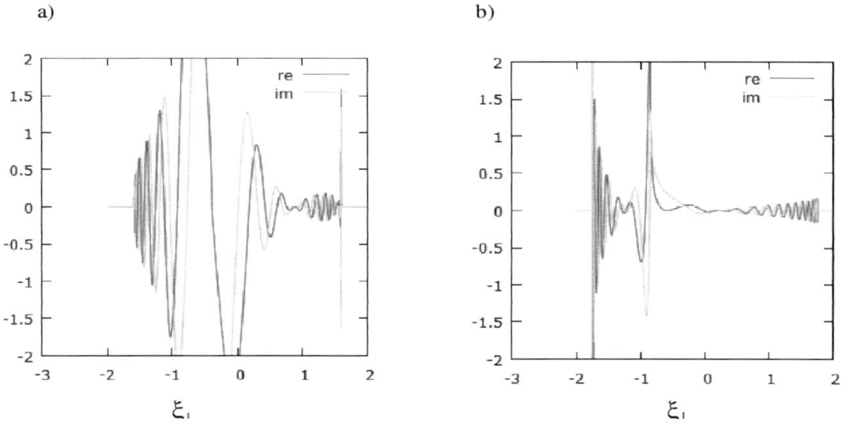

Figure 2. a) Integrand for TV-TV incidence-scatter, $\phi^{\mathrm{i}} = 60°$, $\theta^{\mathrm{i}} = 76°$, $\theta^{\mathrm{d}} = 76°$, $\theta^{\mathrm{sb}} = 77°$. Here $\xi_{st} = -0.398$, $\xi_{br} = -0.4$, $\xi_p = -0.4$, $\gamma_T(0) = 1.6$. b) Integrand for L-TH incidence-scatter, $\phi^{\mathrm{i}} = 60°$, $\theta^{\mathrm{i}} = 60°$, $\theta^{\mathrm{d}} = 70°$. $\xi_{st} = -0.6$, $\xi_{br} = -0.9$, $\xi_p = -0.4$, $\gamma_T(0) = 1.8$. $a = 2\pi \times 10$ wavelengths. Key: red - real part of the integrand, green - imaginary part.

4.2. Standard Simpson's rule

The standard Simpson's rule for a function $H(\xi)$ and the integration interval specified above is

$$\int_{\xi_-}^{\xi_+} H(\xi)\, d\xi \approx \frac{h}{3}\left[H(\xi_-) + H(\xi_+) + 2\sum_{j=1}^{n/2-1} H(\xi_{2j}) + 4\sum_{j=1}^{n/2} H(\xi_{2j-1})\right], \tag{14}$$

with the step size $h = (\xi_+ - \xi_-)/n$ and nodes $\xi_j = \xi_- + j \cdot h$.

4.3. Variable step Simpson's rule

As mentioned above, as we approach the accumulation points, the number of cycles in our highly oscillating integrands increases without bound. To accommodate this behaviour we implement a variable step version of Simpson's rule. Namely, we halve I_δ, and then every new subinterval, repeating the process until $2(n+1)$ subintervals have been generated, with n specified by the user. For

Journal of Physics: Conference Series **498** (2014) 012010 doi:10.1088/1742-6596/498/1/012010

every subinterval, we estimate the length of the smallest cycle of the integrand there. The estimate is given by $2\pi/f'(\xi_1)$ as calculated at the point of the highest oscillation. In accordance with the usual numerical practice, each cycle should be covered by at least $n_{eval} \simeq 10$ nodes. When dealing with subintervals, which lie close to the accumulation points we further multiply the number of nodes there by $1/\gamma_T(\xi_1)$.

4.4. Regularized Simpson's rule

When an integrand $H(\xi)$ contains a pole (say, the geometric pole $\xi_p = -\xi_1^\alpha$, see (10) we can write it in the form $H(\xi) = G(\xi)/(\xi - \xi_p)$ and regularise the integral as follows:

$$\int_{\xi_-}^{\xi_+} \frac{G(\xi)}{\xi - \xi_p} d\xi = \int_{\xi_-}^{\xi_+} \frac{G(\xi) - G(\xi_p)}{\xi - \xi_p} dx + G(\xi_p) \cdot \ln\left(\frac{\xi_+ - \xi_p}{\xi_p - \xi_-}\right). \tag{15}$$

This recipe fails when the pole is close to a node. When it is we shift the whole integration interval so that the pole lies half-way between two nodes.

4.5. Contribution of the pole

The integral (13) can also be seen as

$$\mathcal{I}_\infty = \int_{-\infty}^{\infty} H(\xi)\, e^{iaf(\xi)}\, dx = \left[p.v. \int_{-\infty}^{\infty} + \int_C\right] H(\xi)\, e^{iaf(\xi)}\, d\xi, \tag{16}$$

that is, the sum of the Cauchy Principal Value integral (evaluated using the regularised Simpson's rule) and $\pi i G(-\xi_1^\alpha)$, half residue, the integral over a small half-circle passing under the pole $-\xi_1^\alpha$. Let us justify this choice of the interation path: in the vicinity of the pole the phase can be represented as $f(\xi) \approx f(\xi_p) + f'(\xi_p)(\xi - \xi_p)$. This approximation works when the integrand is fast-oscillating, leading to cancellations outside the neighbourhood of the stationary point. Therefore we can extend the limits of integration beyond the limits of validity of the above Taylor expansion. Taking into account that the phase contains one maximum point ($f''(x_{st}) < 0$), it can be justified more rigorously by introducing a new variable that allows to represent the phase as a quadratic [7]. One way to evaluate the resulting integral is by closing the integration contour in the upper ξ_1 half-plane. This choice is dictated by the radiation condition at infinity: when the pole ξ_p is isolated and $f'(\xi_p) > 0$ the Jordan Lemma can be applied to evaluate its contribution to give the outgoing reflected wave. When $f'(\xi_p) < 0$ the Jordan Lemma is not applicable and the pole makes no contribution. Since the stationary point is a maximum, $f'(\xi_p) > 0$ when $\xi_p < \xi_{st}$.

When wishing to describe the diffracted field only and $\xi_p < \xi_{st}$ the full residue, $2\pi i G(-\xi_1^\alpha)$ should be subtracted from the result, thus removing the asymptotic limit to the reflected field [6], see Fig. 3.

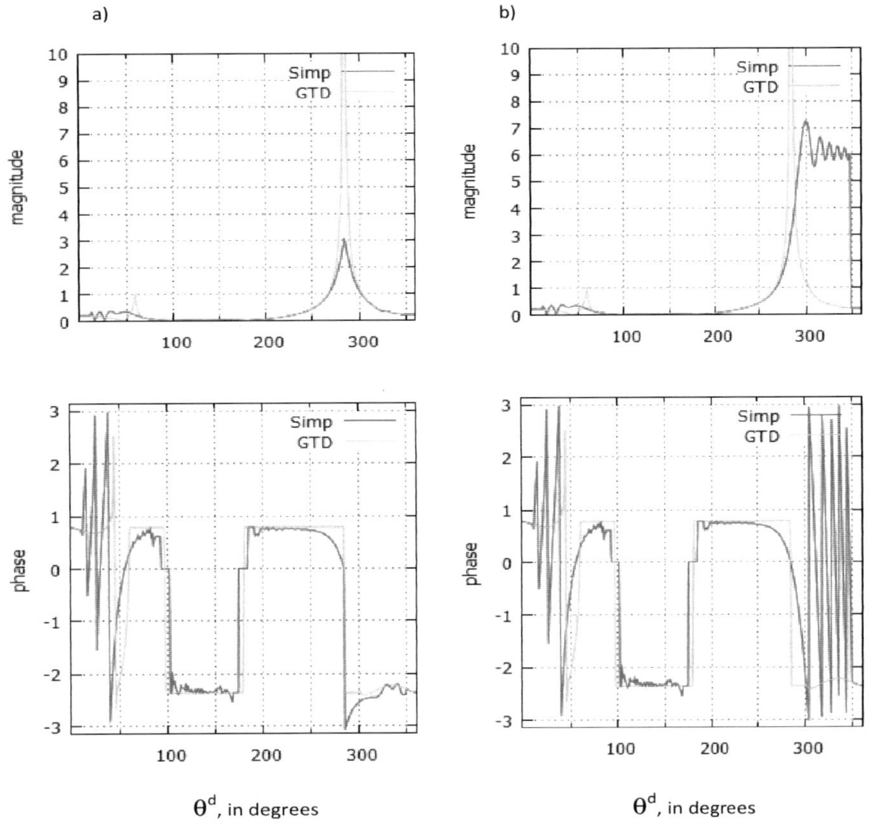

Figure 3. L-TH incidence-scatter, $\phi^{\mathrm{i}} = 60°$, $\theta^{\mathrm{i}} = 60°$, critical angle $\theta^{\mathrm{cr}} = 61°$ and shadow boundary $\theta^{\mathrm{sb}} = 76°$. Diffraction coefficient for $a = 2\pi \times 10$. Both plots show the full numerical model (Simp) versus the GTD calculation (GTD). a) Full numerical model with the pole contribution removed b) Full numerical model with the pole contribution.

4.6. Contribution of the branch point

When evaluating the transverse scattered field the third component of each $\mathbf{v}^{\beta}$ in (10) contains the factor $\gamma^{L}(\xi_1)$ and its domain contains a branch cut that runs along the integration contour from the branch point $\xi_{br} = -\gamma_L$ to the left.

When the branch point ξ_{br} is isolated or coalescing with the stationary point and $f'(\xi_{br}) \geq 0$ the Jordan Lemma can be applied to establish that the branch point makes no contribution. When $f'(\xi_{br}) < 0$ the Jordan Lemma is inapplicable. Thus, when $f'(\xi_{br}) < 0$, even when integrating over the upper side of the cut, the neighbourhood of the branch point makes a contribution. This contribution is known as the head wave. Since the stationary point is a maximum, $f'(\xi_{br}) < 0$ (and therefore, the head wave contribution) is non-zero when $\xi_{br} > \xi_{st}$. The situation becomes more complicated when the branch point

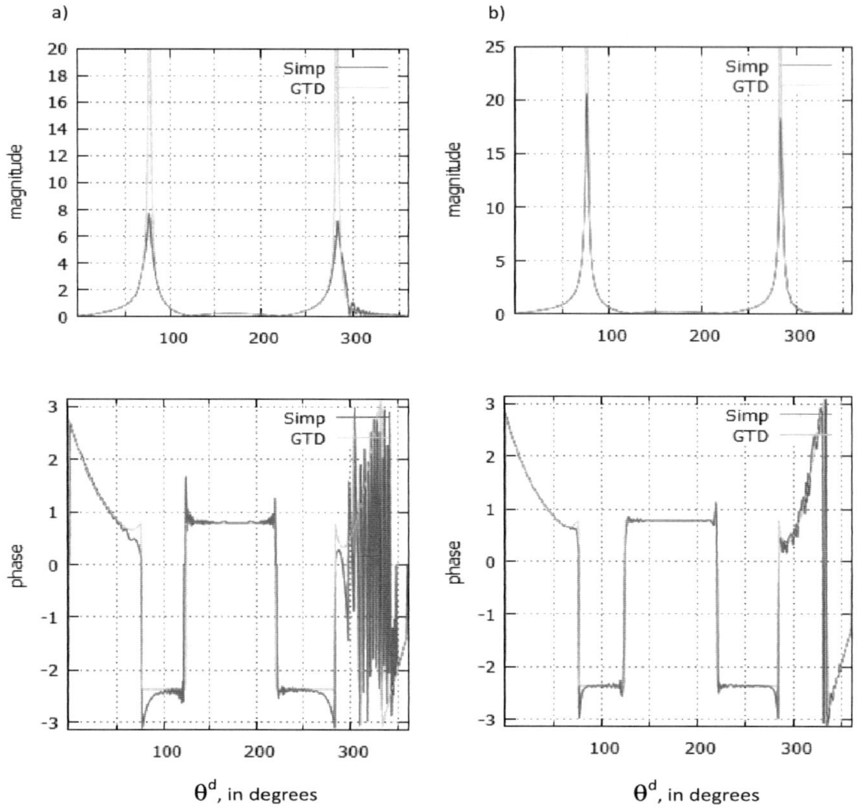

Figure 4. Comparison of the full numerical model (designated Simp) and GTD for the TV-TV incidence-scatter; $\phi^{\rm i} = 60°$, $\theta^{\rm i} = 77°$, $\theta^{\rm cr} = 76°$, shadow boundary $\theta^{\rm sb} = 77°$. a) $a = 2\pi \times 10$, b) $a = 2\pi \times 400$.

coalecses not only with the stationary point but with the pole as well. When $f'(\xi_p) < 0$ and $f'(\xi_{br}) > 0$ and integration is carried out over the upper side of the cut (equivalent to closing the contour in the upper half-plane) neither the pole no branch point make a contribution.

It follows that when the integral (10) is evaluated along the upper side of the cut the results can be contaminated with the contribution of the head wave.

5. Numerical Results and Discussion

Our code is specifically designed to calculate the scattering coefficient when a stationary point coalesces with the branch point $\xi_{st} \simeq \xi_{br}$ or else when all three critical points ξ_{st}, ξ_{br} and ξ_p coalesce. For this reason it employs the modified Simpson's method only when the stationary and branch point are coalescing, $\xi_{st} \simeq \xi_{br}$, relying on the classical GTD otherwise. We refer to it as the *mixed GTD/numerical code.* We say that the stationary point and branch point are coalescing when the phase (11) satisfies condition $|a \cdot f(\xi_{st}) - a \cdot f(\xi_{br})| < 6\pi$.

For validation purposes, below we use only the *full numerical model*, which utilises the Simpson's method at all observation angles. At observation angles near grazing, where we have 0°, 180°, the calculations are always carried out using GTD. Everywhere we choose $\delta = 0.1$, $n_{eval} = 10$ and $2n = 10$. Let us illustrate the performance the code in the region of interference of the reflected, diffracted and head waves. In all examples below $c_P = 5890\, m \cdot s^{-1}$ and $c_S = 3210\, m \cdot s^{-1}$ as in mild steel.

In Fig. 4 we compare performance of the full numerical model with GTD in the transition region where the head wave interacts with both a diffracted wave and reflected wave. The figure shows that at critical angles, at smaller distances from the diffraction point the numerical method predicts a smaller value of the scattering coefficient than GTD. Other authors, see e.g. [8] and [9], report similar behaviour for diffraction by a finite strip. As explained in [9], for the finite strips the smoothing of the critical spikes in the diffracted field is due to interference between the head waves and diffracted waves generated by the strip edges as well as the reflected Rayleigh waves. We believe that a similar phenomenon explains our results: at smaller distances where the head wave is still significant there is destructive interference between it, diffracted wave and maybe the reflected wave. This conclusion is supported further by Fig. 4b) which demonstrates that for larger values of the far field parameter, the full numerical method produces the scattering coefficient with the peaks at critical angles, which are higher than for $a = 2\pi \times 10$. They are smaller than those in GTD coefficients, since the latter fails in penumbras and in the reported configuration shadow boundaries and critical rays coalesce. Other numerical experiments carried out where GTD applies, away from the shadow boundary, confirm that the GTD peaks at critical angles are correct: they are the same as the peaks obtained with the full numerical model.

References

[1] Achenbach J D and Gautesen A K 1986 *Edge diffraction in Acoustics and Elastodynamics* (Low and High Frequency Asymptotics, vol 2) ed V K Varadan and V V Varadan (Amsterdam: North-Holland) chapter 4 pp 335-99

[2] Achenbach A, Gautesen A and McMaken H 1979 *Ray Methods for Waves in Elastic Solids, With Applications to Scattering by Cracks* (New York: Pitman)

[3] Darmon A, Leymarie N, Chatillon S and MahautS 2009 *Modelling of scattering of ultrasounds by flaws for NDT* (Ultrasonic Wave Propagation in Non Homogeneous Media) (Springer Proceedings in Physics vol 128) (Berlin: Springer), pp 61-71

[4] Gautesen A 2010 *Wave Motion* **47**(5) 327

[5] Mahaut S, Darmon M, Chatillon S, Jenson F and Calmon P 2009 *Insight*, **51**(2), pp 78-81(4)

[6] Borovikov V A 1994 *Uniform Stationary Phase Method* (London: IEE)

[7] Bleistein N and Handelsman R 1975 *Asymptotic Expansions of Integrals* (New York: Dover)

[8] Chapman R K 1982 *Ultrasonic scattering from smooth flat cracks: An elastodynamic Kirchhoff diffraction theory* (North Western Region NDT Applications Centre: CEGB Report NWR/SSD/82/0059/R)

[9] Keogh P S 1985 *The Quaterly Journal of Mechanics and Applied Mathematics* **38**(2) pp 205-232

Journal of Physics: Conference Series **498** (2014) 012011 doi:10.1088/1742-6596/498/1/012011

Modelling of waves propagation on irregular surfaces using ray tracing and GTD approaches: Application to head waves simulation in TOFD inspections for NDT

Adrien Ferrand[1], Michel Darmon[1], Sylvain Chatillon[1], Marc Deschamps[2]

[1] CEA, LIST, Department Imaging and Simulation for NDT, 91191 Gif-sur-Yvette cedex, France.
[2] Univ. Bordeaux, I2M–APy, UMR 5295, 33400 Talence, France.

Email : michel.darmon@cea.fr

Abstract. The Time of Flight Diffraction (TOFD) technique is a classical ultrasonic method used in ultrasonic non-destructive evaluation, which allows a precise positioning and a quantitative size evaluation of cracks in the inspected material. Among the typical phenomena arising in the current TOFD inspection, the so-called "head wave" is the first contribution reaching the receiver. The head wave propagation on a planar interface is well known and identified as a critical refraction taking place on the material surface. On irregular surfaces, it has been shown that the head wave results from the melting of surface and bulk waves mechanisms and that surface irregularities are responsible for numerous diffractions of the incident head wave. To simulate such behaviour, a model has been developed using a ray tracing technique based on time of flight minimization (generalized Fermat's principle). It enables the calculation of the ray path and the corresponding time of flight of all waves propagating in the material, including the head wave. To obtain a complete propagation model for these waves (both trajectory and amplitude), the integration of Geometrical Theory of Diffraction (GTD) models is currently performed by coupling them with the ray-based approach discussed above.

1. Introduction

Among the various NDE inspections techniques, the Time of Flight Diffraction (TOFD) technique allows a precise and quantitative evaluation of surface breaking or embedded cracks. This technique uses two ultrasonic probes which are mechanically associated and remote from each other. In a contact inspection, the two probes are positioned on the surface of the inspected specimen using a wedge to ensure the impedance adaptation between the probes and the specimen. The first probe emits an ultrasonic pulse in the specimen, and the second probe receives the diffracted signals arising from the specimen. In Fig. 1, a schematic example of a TOFD inspection on a specimen with a planar surface is provided.

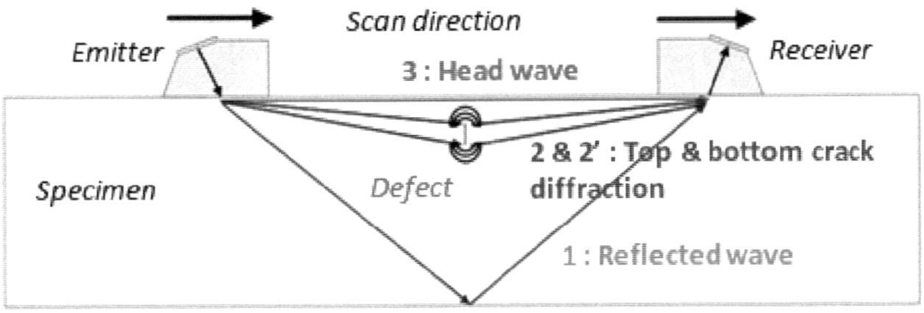

Figure 1. TOFD inspection on a specimen with a planar surface.

Typical waves received by the second probe and their ray representation are shown in Fig. 1. Among these contributions, specular reflections on the bottom of the specimen and diffraction on the defect edges are received. But a wave, represented in Fig. 1 and named the head wave, is detected chronologically before any other. This wave, generated by a critical refraction at the angle between the wedge and the specimen, propagates along the entry surface and finally give rise to a critical radiation from the specimen to the receiver wedge. The principle of head wave propagation along a planar surface separating two fluid media of respective density (and L-wave velocity ( ) is summed up in Fig. 2. The L-wave emitted from the source point is critically refracted at the point A, and critically radiated from the surface at the point B towards the observation point.

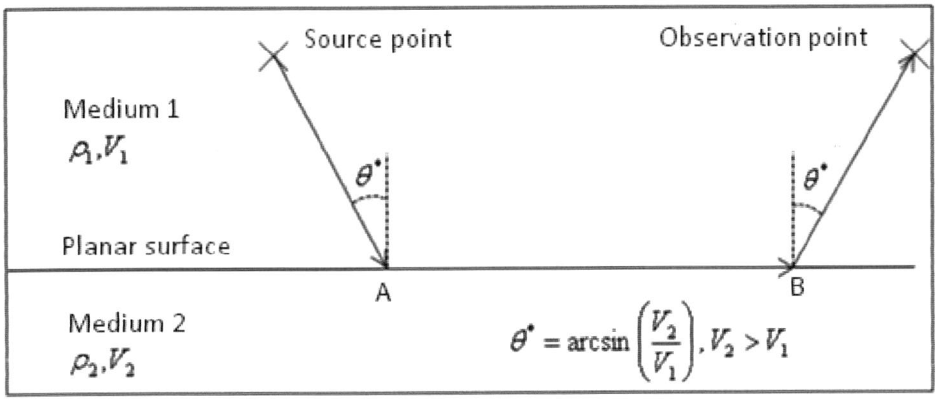

Figure 2. Head wave propagation principle along a planar interface.

The head wave has been historically studied in geophysics. Indeed, seismic inspections often involve stratified media which allow head wave generation on the media interfaces [1]. Analytical models have been developed in the case of stratified media separated by planar interfaces [1], [2], [3], [4]. One of these models [4] is based on a ray approach and asymptotical expansion of the propagation solution: using the hypothesis of a high frequency point source, the head wave propagation can be modeled using a ray approach and can be analytically expressed until the observation point. Furthermore, analytical modelling of the head wave propagation on irregular surfaces has also been proposed in several seismic studies. Indeed, using an integral representation, the received head wave signal has been evaluated for rough surfaces or slightly curved surfaces [5,6] and for spherically symmetric radially heterogeneous media [7]. Some numerical studies [8] have also been performed concerning more generally irregular surfaces and have shown some effects of the irregular surface on the head wave amplitude and time of flight.The objective of this study is to extend the head wave modeling from the planar interfaces case to the irregular interfaces case using a ray approach in the NDE domain.

The irregular interfaces studied in this work are the scoured surfaces of metal specimens. Scoured surfaces are the result of a specimen repair process. When a defect is detected by a first TOFD inspection near the specimen surface, the matter around the defect is removed in view to be replaced by clean matter. Before the replacement, a new TOFD inspection is carried out on the resulting scoured surface in order to ensure that the entire defect has been removed. During the new TOFD inspection, the measured head wave signal is modified compared to the first TOFD inspection. To understand the differences between the two head wave signals, a test specimen has been designed. This specimen shown in Fig. 3a has a surface divided in two parts: one is planar, the other is scoured. A TOFD inspection is made along the surface, and the resulting B-scan is shown in Fig. 3b.

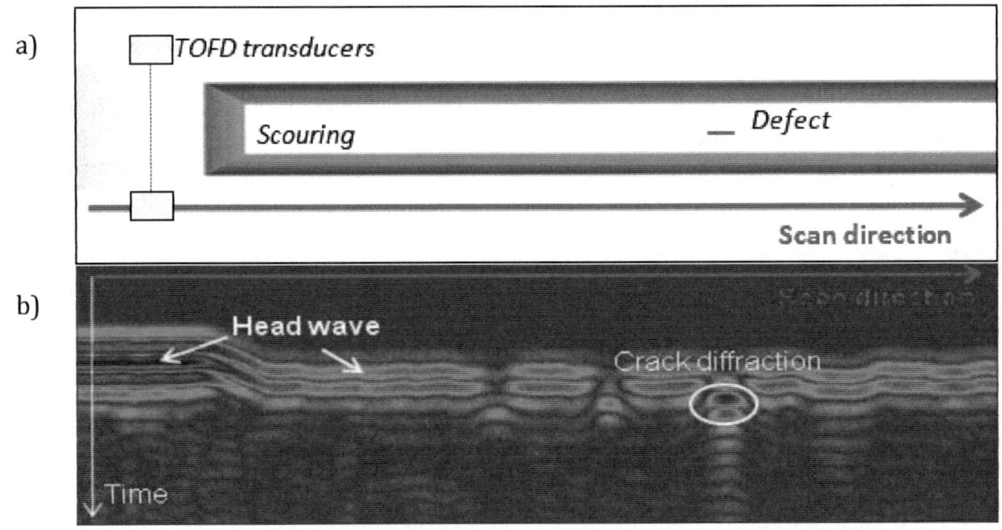

Figure 3. a) Inspection configuration of the test specimen. b) Measured B-scan on both the planar part and the scoured part of the surface

The B-scan obtained during the surface inspection from the planar part to the scoured one represented in Fig. 3, focuses on the effect of the surface irregularity on the head wave signal. In fact, the time of flight of the head wave signal shows that the head wave is delayed when the TOFD transducers reach the scoured part of the surface, which indicates that the head wave propagation is modified. Moreover the head wave signal amplitude measured on the scoured part significantly decreases, (around 20dB) compared with that obtained on the planar part. The modeling of this variation implies the study of head wave scattering by surface irregularities.

The description of the modeling developed in this article is divided in three steps. Firstly, a new ray tracing algorithm has been developed in order to calculate the ray path of the head wave on irregular surfaces (section 2). Secondly, an asymptotic model is proposed to obtain the head wave amplitude on a particular irregular surface (section 3). Finally, some results of a complete model (ray path and amplitude) of the head wave propagation are presented (section 4) for the previously studied surface.

2. Ray modelling of head wave propagation near irregular surfaces

This section is devoted to the description of the developed generic ray tracing tool which allows calculating the head wave path. The head wave scattering on surface irregularities of a specimen is simulated using the hybrid technique CIVA/Athena [9]. This technique associates the analytical CIVA pencil method for beam modelling and Athena finite elements method (FEM) for beam to scatterer interaction modelling. The results obtained are then considered as a reference to carry out a phenomenological analysis on the scattering effects of the irregular surface. Such an analysis shows that these scattering effects can be interpreted in terms of diffracted rays accordingly to the GTD (Geometrical Theory of Diffraction) principles [10]. The generic ray tracing tool has been developed based on the GTD principles and then validated by comparing its results on the wave ray path with the reference results of CIVA/Athena simulations.

2.1. Interpretation of head wave path near irregular surfaces

In order to interpret the head wave propagation near irregular surfaces, a TOFD inspection simulation has been carried out using the hybrid technique CIVA/Athena on a specimen presenting two flat slopes on its entry surface: these slopes form a corner along the incident surface and correspond to the surface irregularity. The inspection configuration is described Fig. 4.

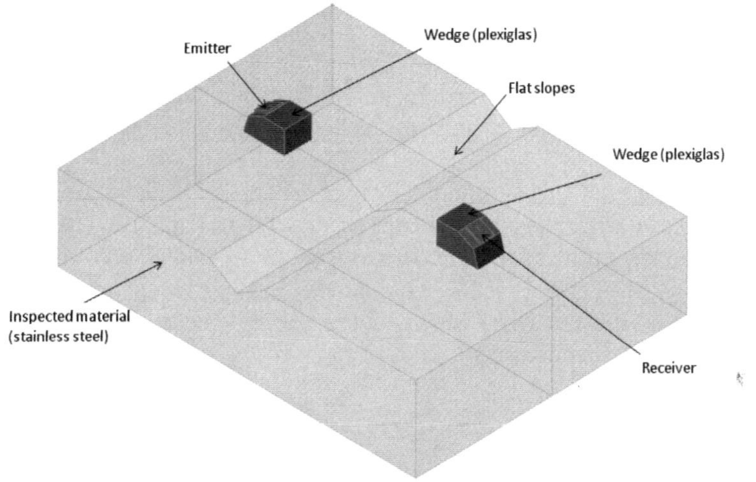

Figure 4. Inspection configuration for the interpretation of the head wave path

The instantaneous elastodynamic field is calculated near the surface irregularities and is represented in Fig. 5. The area near the surface irregularity (a wedge) appears to be divided in two parts: the first one is the specimen area illuminated by the ultrasonic emitted beam and the second one represents the shadow created by the irregularity part. The two parts are separated by the so-called shadow/light boundary whose location is defined by the shape of the surface irregularity and by the emitting transducer position with respect to the surface irregularity. Fig. 5 shows that an elastodynamic field exists in the shadowed area and is damped compared to the one in the insonified area.

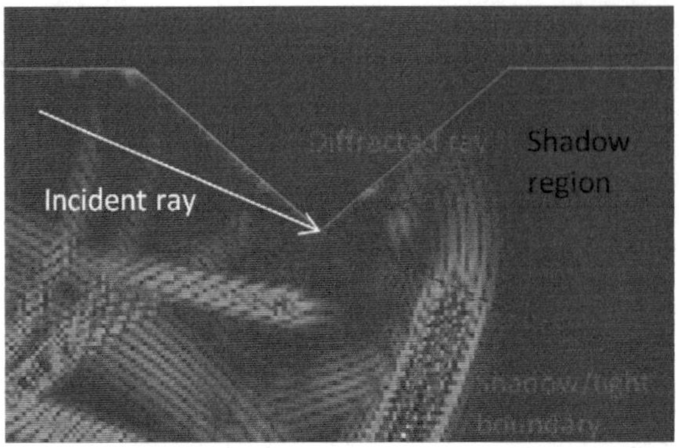

Figure 5. Simulated elastodynamic field around the surface irregularity

In fact, when reached by the incident field, the surface irregularity generates diffraction. The observed diffraction phenomenon corresponds to the canonical problem of wedge diffraction which can notably be predicted by GTD: the wave propagation in the shadowed area can be represented in this case by rays diffracted from the wedge. Moreover the diffracted field generated by the surface irregularities propagates in the specimen and the corresponding wave front reaches the receiver so as to generate the head wave signal received during the TOFD inspection.

Consequently the results of this simulation show that the head wave path is that represented in Fig. 6. The incident ray generated by the contact emitter reaches the entry specimen surface and creates a refracted ray in the specimen. This refracted ray is diffracted on the surface corner into a diffracted ray as predicted by GTD. Then the diffracted ray propagates in the specimen and is then refracted at the interface between the entry specimen surface and the receiver wedge and reaches the receiver crystal.

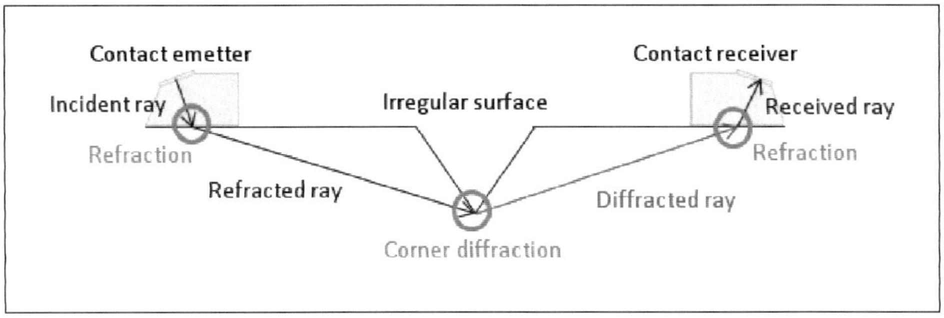

Figure 6. Head wave ray path along the corner surface irregularity

In conclusion the ray approach from the GTD principles is able to interpret the head wave propagation near irregular surfaces, and has been retained to develop the head wave modeling.

2.2. Development of a generic ray tracing tool

A generic ray tracing tool has been developed in order to calculate the head wave path as described in section 3.1. This generic ray tracing tool uses one basic GTD principle, the Generalized Fermat's Principle, whereby a ray path must satisfy the principle of stationary time of flight in order to represent the wave propagation. The Generic Ray Tracing Tool (GRTT) accounts of diffraction from the surface irregularities of the inspected specimen and is able to simulate paths of all propagating waves including head waves. The algorithm functionalities and description are the subject of a patent filing [11].

To validate the head wave path calculation obtained by the generic ray tracing tool, similar simulations as those performed in section 3.1 have been carried out on two types of surface irregularities: the first one concerns the same specimen as in Fig. 4 which presents two flat slopes on the entry surface, and in the second one shown in Fig. 7b the flat slopes are replaced by cylindrical ones. The elastodynamic field near the surface irregularities and the head wave front is simulated by both GRTT and the FEM hybrid method. Ray paths implying a diffraction of the head wave from the surface irregularities have been calculated in the specimen bulk with the generic ray tracing tool in order to build the theoretical head wave front showing the diffraction effect also predicted by GTD. Thus the GRTT validation consists in comparing the simulated head wave fronts calculated by both hybrid technique CIVA/Athena and GRTT. The comparison is shown in Fig. 7 for the two specimens.

a)

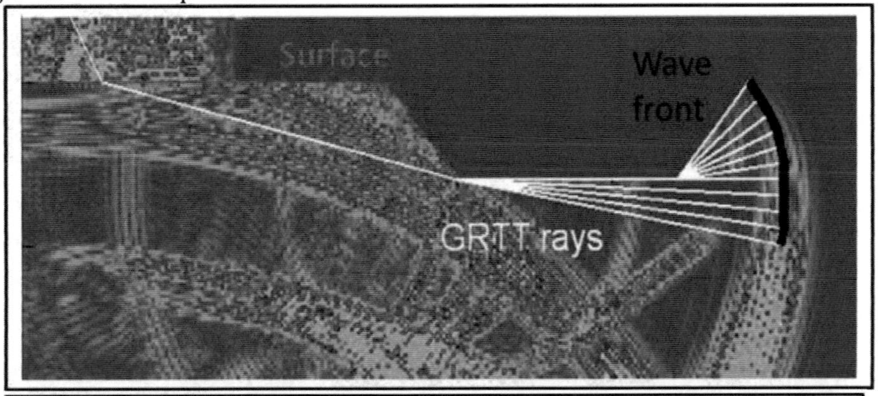

b)

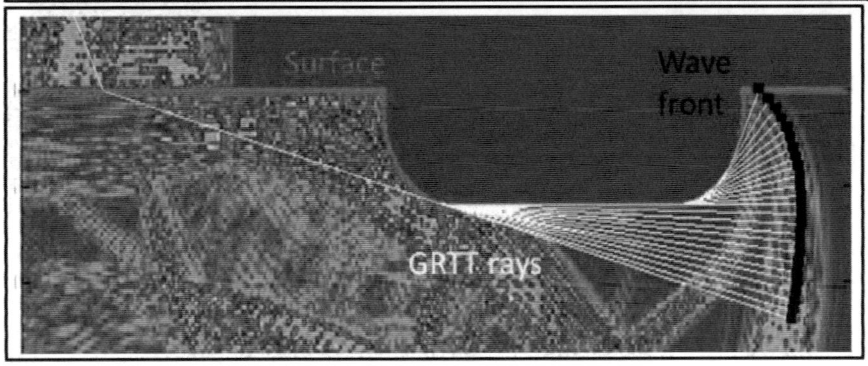

Figure 7. Rays and wave front calculated by GRTT compared to the elastodynamic field (on colours maps) calculated by CIVA/Athena for : a) a specimen with corners, b) a specimen with curved surfaces

As observed in Fig. 7, the theoretical head wave fronts calculated in the two specimens are in good agreement with the diffracted wave fronts obtained with the hybrid technique CIVA/Athena. The comparisons done in the bulk specimen validate the GRTT ray approach used to simulate the diffractions effects responsible of the head wave propagation near irregular surfaces.

3. Development of GTD models for head waves amplitude on curved surface irregularities

The GRTT described in section 3 provides the ray and time of flight of the head wave. Now the calculation of the head wave amplitude with GTD models using the knowledge of the ray path is described. To this aim, the GTD approach is used: it refers to mathematical methods of asymptotical developments [8,9] which allow a complete calculation of the head wave amplitude by the study step by step of the diffraction effects along the head wave ray path.

3.1. Principle of amplitude calculation along a ray-based model

Ray path calculated by GRTT is composed of one or more diffracted or specular rays connected to each other. Each ray composing the ray path must be studied independently to calculate the total amplitude of the wave propagating along the ray path. Indeed the GTD approach expresses the amplitude carried by a ray as follows: for a ray of length *r*, the displacement on the ray extremity is given by the amplitude model:

$$u_d(r) = \mathrm{A}_{source} \exp(i\psi_{source}) \;\; D \;\; L(r) \;\; \exp(ikr) \tag{1}$$

where $\mathrm{A}_{source} \exp(i\psi_{source})$ is the amplitude carried by the incident ray, D is the diffraction coefficient depending on the nature of the effect responsible of the ray launching, $L(r)$ is the divergence factor of the ray, and $\exp(ikr)$ is the phase factor of the ray (k being the wave number). The previous formula shows that GTD expresses the amplitude of a ray as the product of the incident field $\mathrm{A}_{source} \exp(i\psi_{source})$ and of a scattering term $D \;\; L(r) \;\; \exp(ikr)$ which depends on the nature of the ray (diffracted in the bulk from an irregularity, creeping...) and on its geometrical characteristics.

The approach for a single ray is generalized in order to obtain the amplitude of a wave propagating along a ray path with the following algorithmic principle. An example is given on the Fig. 8. Firstly, the GRTT is used to generate the ray path of the head wave between the known source and observation points. Then the wave propagation is studied along the ray path in order to detect all the interactions between the wave and the surface. Each interaction i generates the ray i with the corresponding amplitude model (with the associated scattered displacement $u_d^{(i)}(r_i)$) which depends on the nature of the interaction, the geometrical parameters of the ray i and the incident field $u_d^{(i-1)}(r_{i-1})$ coming from the ray $i-1$. Finally the total amplitude of the wave propagating along the ray path on the observation point is given by the concatenation of all the amplitude models:

$$u(P_{observation}) = f(u_d^{(1)}(r_1), u_d^{(2)}(r_2), \ldots, u_d^{(n)}(r_n)) \tag{2}$$

$$u(P_{observation}) = u(\mathrm{P}_{source}) \prod_{i=1}^{n} D_i \left[\sqrt{d\Sigma(r_i^{'}) / d\Sigma(r_i)} \right] \exp(ikr_i) \tag{3}$$

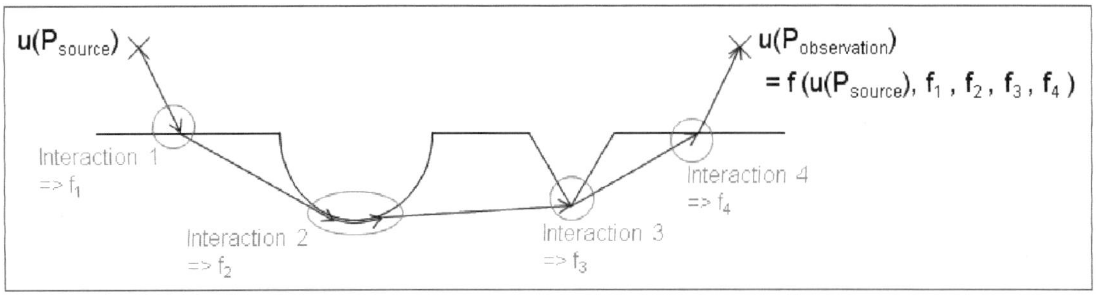

Figure 8. Example of an amplitude calculation along a ray path calculated by GRTT

The amplitude calculation of a wave along a ray path is then made by the application of adapted amplitude models on each ray composing the ray path. These models are asymptotic ones (GTD) and one model will now be presented for a specific diffraction effect occurring in TOFD inspection near irregular surfaces: the diffraction of the head wave on a curved surface irregularity.

3.2. Principle of amplitude calculation along a ray-based model

The head wave ray path calculated by GRTT in specimens with curved surfaces irregularities has been presented in Fig. 7b. It shows a particular diffraction along the irregularities which is presented in Fig. 9 for the case of a cylindrical one with a point source S and an observation point located on Q. It appears that the ray path integrates a bulk ray SP_1 which is tangentially incident to the curved surface irregularity and launch a surface ray P_1P_2 along this irregularity. After the surface propagation, this ray generates a diffracted bulk ray P_2Q tangentially to the irregularity in the shadow region formed by the surface. The ray P_1P_2 is predicted by the GTD and is called a creeping ray.

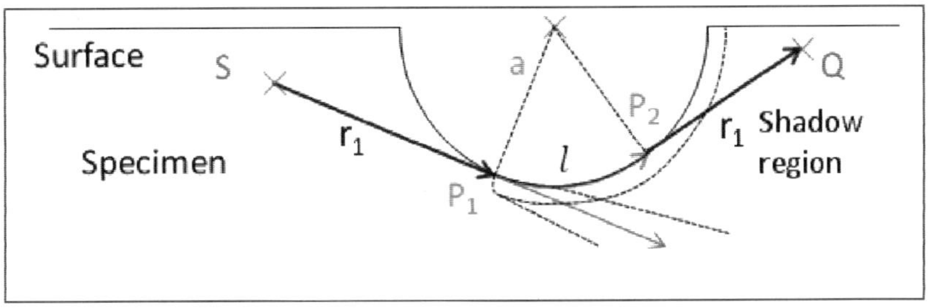

Figure 9. Creeping ray path along an cylindrical surface irregularity

The amplitude of the wave propagating along the path SP_1P_2Q results from the diffraction of the creeping ray P_1P_2 in the shadow region. The corresponding chosen creeping ray amplitude model derivates from the canonical problem of plane wave diffraction from a soft cylinder of radius a using an asymptotic expression of the solution only valid for a scalar wave at high frequency in the shadow region of the cylinder (mode conversions of elastic case are neglected). For a line source (2D approximation) S emitting at the wave number k, this model [13] gives the following amplitude:

$$u(Q) = D_S \frac{e^{ikr_1}}{\sqrt{r_1}} \sum_{n=0}^{\infty} D_{sr,cr}^{(n)} D_{cr,sr}^{(n)} e^{i\frac{\nu_n}{a}l} D_S \frac{e^{ikr_2}}{\sqrt{r_2}} \quad (4)$$

With D_S the dyadic source coefficient equal to $e^{i\pi/4}/\sqrt{8\pi k}$, the first term $D_S \frac{e^{ikr_1}}{\sqrt{r_1}}$ is the incident ray SP_1 amplitude and the last term $D_S \frac{e^{ikr_2}}{\sqrt{r_2}}$ is the diffracted ray P_1Q amplitude generated by the creeping ray. The term $\sum_{n=0}^{\infty} D_{sr,cr}^{(n)} D_{cr,sr}^{(n)} e^{i\frac{\nu_n}{a}l}$ is the asymptotic expansion of the creeping ray model and includes the diffraction effect on the curved surface. Indeed, $D_{sr,cr}^{(n)}$ and

$D^{(n)}_{cr,sr}$ are respectively the attachment and detachment coefficients on P_1 and P_2 of the creeping ray and can be expressed using first order Airy functions:

$$D^{(n)}_{sr,cr} D^{(n)}_{cr,sr} = D_n^2 = \frac{2^{2/3} e^{-i\pi/6} (ka)^{1/3}}{\left[A_i^{'}(-\bar{\eta}_n) \right]^2}, \tag{5}$$

where $\bar{\eta}_n$ are the zeros of the Airy function. Furthermore the ν_n term of the asymptotic expansion is the attenuation term of the creeping ray which represents the energy loss due to the diffraction in the bulk during the creeping ray propagation. ν_n is expressed using the zeros $\bar{\eta}_n$ of the Airy function:

$$\nu_n = ka + \left(\frac{ka}{2} \right)^{1/3} e^{i\pi/3} \bar{\eta}_n \tag{6}$$

4. Simulated results

The creeping ray amplitude model has been evaluated by a parametric study of the asymptotic expansion $\sum_{n=0}^{\infty} D^{(n)}_{sr,cr} D^{(n)}_{cr,sr} e^{i\frac{\nu_n}{a}l}$ presented in section 3.1. The two significant parameters of the model are the radius of curvature a of the curved irregularity and the frequency of the head wave. For a fixed creeping ray length of 10 mm, the Fig. 10 presents the results of the parametric study for typical values of a and f encountered in TOFD inspections.

a)

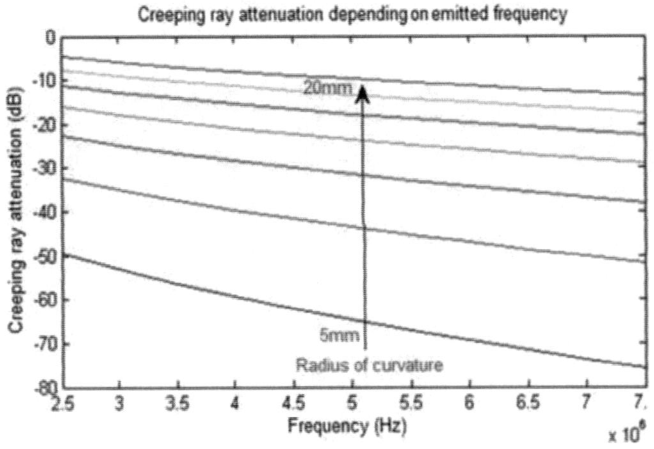

b)

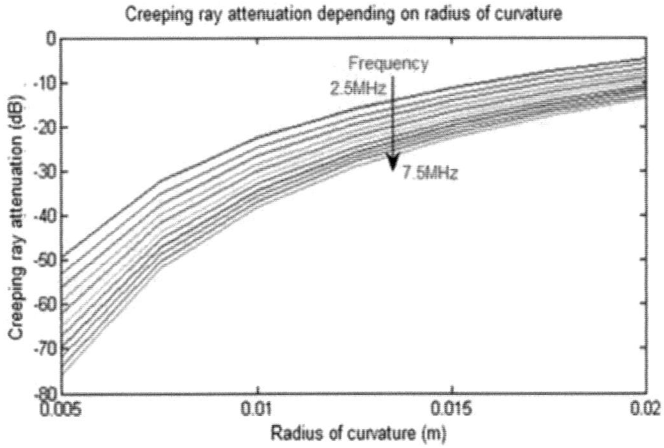

Figure 10. Results for the parametrical study of the asymptotic expansion : wave attenuation over frequency (a) and radius of curvature (b)

The parametric study shows that the wave propagation along a creeping ray implies a strong attenuation of its amplitude.

The amplitude calculation approach of section 3.1 has been followed to integrate the creeping ray amplitude model in the GRTT and to obtain a complete simulation of head wave fronts propagating along curved surfaces irregularities. Some simulations have been then carried out on the specimen presented in Fig. 8 (for a radius curvature of the irregularity of 10 mm and a harmonic line source of frequency 5MHz): the GTD creeping ray model presented in section 3 has been used in the shadow area formed by the cylindrical surface irregularity and the wave in the insonified area has been simulated using the classical results of Geometrical Acoustics. The calculated wave fronts for different steps of the propagation are shown in Fig. 11. The amplitude carried by the rays is represented using a color code with a reference amplitude equal to the mean amplitude of the front in the specimen before reaching the surface irregularity.

a)

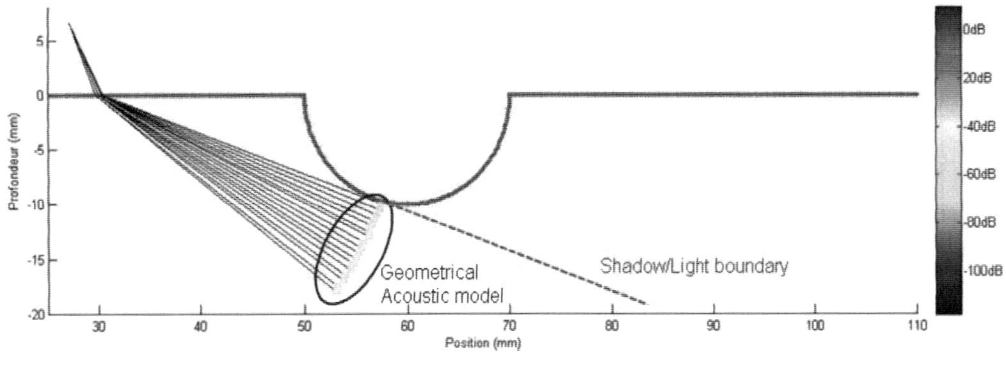

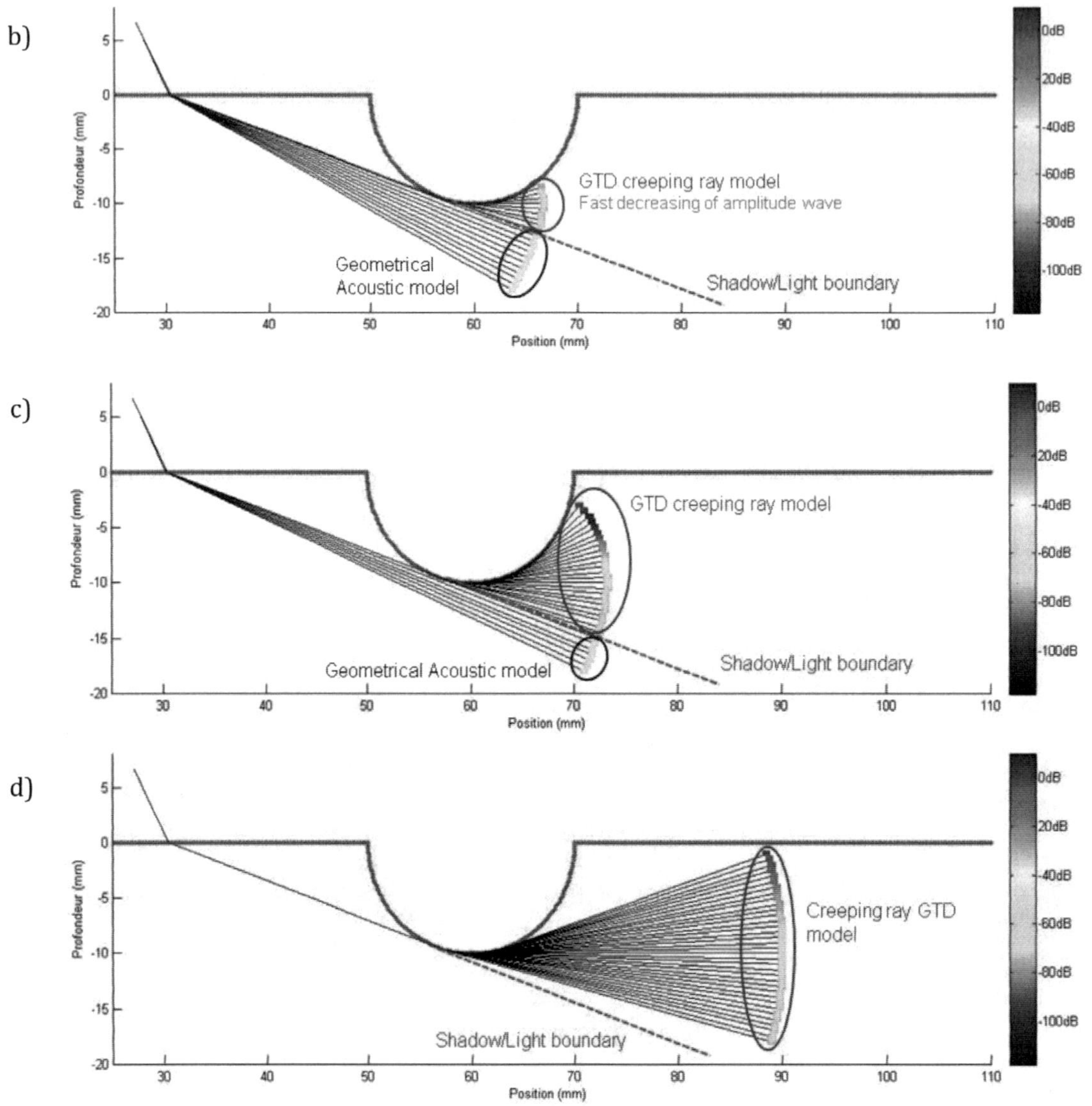

Figure 11. Complete simulation (ray paths and amplitude) of the wave front by the GRTT at different steps of the propagation (a, b, c, d)

The curved surface irregularity forms a shadow area (Fig. 11a) in which the head wave is diffracted from the irregularity. In the shadow area, the creeping ray model is applied (Fig. 11b, 11c). The resulting simulation shows that the head wave front in the shadow area is strongly attenuated during its propagation (Fig. 11d). Far from the irregularity, the amplitude attenuation in the shadow zone is of 20dB to 40dB compared to the insonified zone. The complete simulation of the head wave including ray path, time of flight, wave front and amplitude is thus done for the case of cylindrical surface irregularities.

Conclusion

Classical models of head waves along planar surfaces cannot be used to simulate head waves along irregular surfaces in TOFD inspection. Whereas the head wave behavior on planar surfaces postulates a surface propagation, head waves on irregular surfaces are the result of both surface and bulk propagation.

In order to simulate the head wave propagation in such irregular surfaces by a ray approach, a generic ray tracing tool (GRTT) has been developed to model the diffraction phenomena predicted by the Geometrical Ray Theory, arising along the surface irregularities and responsible of the head wave propagation. The GRTT has been validated concerning the ray path and the time of flight of head waves. Moreover specifics GTD models using asymptotic developments have been integrated to calculate the head wave amplitude and then provide a complete simulation of the head wave on cylindrical surface irregularities for TOFD inspections.

References

[1] L. M. Brekhovskikh, *Waves in layered media*. Academic Press, 1960.
[2] L. Cagniard, E. A. Flinn, C. H. Dix, and W. G. Mayer, "Reflection and Refraction of Progressive Seismic Waves," *Phys. Today*, vol. 16, p. 64, 1963.
[3] A. De Hoop, "A modification of Cagniard's method for solving seismic pulse problems," *Appl. Sci. Res. Sect. B*, vol. 8, no. 1, pp. 349–356, 1960.
[4] V. Červený and R. Ravindra, *Theory of seismic head waves*. University of Toronto Press (Toronto), 1971.
[5] I. Lerche and N. Hill, "A mean-field solution of the reflection of a spherical acoustic wave from a rough interface," *J. Math. Phys.*, vol. 26, p. 1420, 1985.
[6] I. Lerche, "On the reflection of acoustic waves from a slightly curved interface," *J. Acoust. Soc. Am.*, vol. 81, p. 611, 1987.
[7] D. P. Hill, "Critically refracted waves in a spherically symmetric radially heterogeneous earth model," *Geophys. J. Int.*, vol. 34, no. 2, pp. 149–177, 1973.
[8] H. Zhou and X. Chen, "Ray path of head waves with irregular interfaces," *Appl. Geophys.*, vol. 7, no. 1, pp. 66–73, 2010.
[9] N. Gengembre, A. Lhémery, R. Omote, T. Fouquet, and A. Schumm, "A semi-analytic-FEM hybrid model for simulating UT configurations involving complicated interactions of waves with defects," in *AIP Conference Proceedings*, 2004, vol. 700, p. 74.
[10] J. B. Keller, "Geometrical theory of diffraction," *JOSA*, vol. 52, no. 2, pp. 116–130, 1962.
[11] Patent filing related to GRTT in progress. Its precise reference will be given in the proofs.
[12] D. Bouche, F. Molinet, and R. Mittra, *Asymptotic methods in electromagnetics*. Springer, 1997.
[13] F. Molinet, *Acoustic High-Frequency Diffraction Theory*. Momentum Press, 2011.

Journal of Physics: Conference Series **498** (2014) 012010 doi:10.1088/1742-6596/498/1/012010

A modal formulation for the propagation of guided waves in straight and curved pipes and the scattering at their junction

M El Bakkali[1], A Lhémery[1], V Baronian[1] and S Grondel[2]

[1]CEA, LIST, F-91191 Gif-sur-Yvette cedex
[2] IEMN-DOAE, UMR CNRS 8520, Université de Valenciennes et du Hainaut Cambrésis, Le Mont Houy, F-59313 Valenciennes Cedex 9

E-mail: alain.lhemery@cea.fr

Abstract. The well-established semi-analytic finite element (SAFE) method for predicting guided wave propagation in straight guides is extended to deal with curved pipes using curvilinear coordinates system, following developments by Treyssède and Laguerre [*J. Sound Vib.* **329** 1702-16 (2010)] for helical wires. The modal solution in the elbow can be obtained and the bi-orthogonality relation known for straight guides still holds. The influence of elbow curvature is exemplified by a short parametrical study. The scattering at the junction between a straight guide and an elbow sharing a common section is then considered. Assuming the modal solutions in the straight part and in the curved one are known, the mode-matching method is specifically derived to compute the scattering matrix which deals with waves arising from one part or the other, that is to say, to predict reflection and transmission coefficients at such a junction. Interestingly, the overall formulation requires a single mesh of the section to compute both the modal solutions and the scattering. The scattering matrix computation makes it necessary to take into account evanescent modes locally at the junction of both parts. Again, few examples of numerical results are shown to illustrate specific phenomena arising in this complex scattering case.

1. Introduction

Elastic guided waves (GW) non-destructive techniques are widely used for in-service inspection of pipeline networks in various industries (nuclear, oil, chemical industries...). However, because of the complexity of scattering phenomena involved when guided waves in a straight pipe interact with specific features such as flaws, welds, supports, elbows, junctions, etc., tools for simulating nondestructive examinations must be developed to help the interpretation of measured signals and to improve testing configurations.

In pipe networks, straight lines coexist with elbows. Guided wave propagation in the former parts is well-known and several computational methods as well as the semi-analytic finite element method (SAFE) or Finite Element Method (FEM) can be used to predict it. In the latter parts, the propagation of guided waves is far less documented [1]. The scattering of guided waves at their junction constitutes an even more difficult problem.

The well-known SAFE method is the most commonly used modeling approach to calculate the GW modal solution in components of various shapes. Compared to the FEM, this technique is less computationally expensive and allows dealing with guides of arbitrary section.

In the present paper, the aim is to establish a numerical model based on the SAFE method able, in a first place, to deal with the curvature effect on the elastic guided waves propagation through a pipe elbow and then, to quantitatively predict the scattering phenomena arising when a guided wave is diffracted at the junction between a straight pipe and an elbow.

The paper is organized as follows. The calculation of the modal solution with the SAFE method in a curved guide and a short parametrical study of the curvature effect are first presented. Then, the simulation of the scattering at the junction between a straight guide and a curved one is carried out by using the mode matching method, where modal solutions in both guides are linked. Finally, examples of numerical results are shown to illustrate specific phenomena arising at this junction.

2. SAFE model in curvilinear coordinates system

2.1. Equation of wave motion

The waveguide considered here is a straight-curved-straight pipe (Fig. 1)

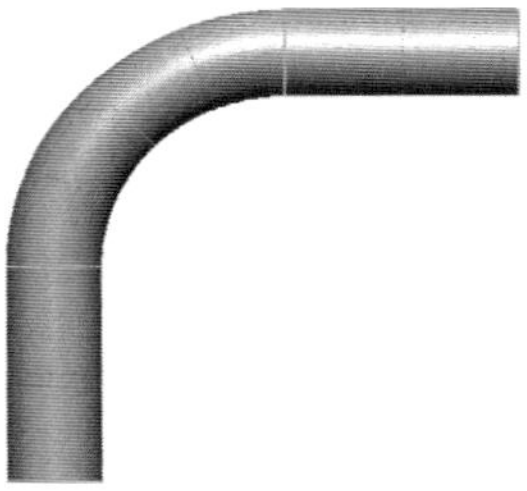

Figure 1. Schematic view of the studied elbow configuration

Our aim is to calculate the scattering matrix at the junction between a straight guide and a curved guide. The first step consists in computing the modal solution in a curved waveguide by means of the SAFE method. The second step consists in computing the scattering matrix at the junction using a mode matching technique.

Guided modes are solutions of the wave equation in an elastic waveguide (pipe in the present case) assumed to have free boundaries and to be homogeneous. The equation of motion is:

$$\begin{cases} -\mathbf{div}\,\sigma(\mathbf{u}) - \omega^2 \rho\,\mathbf{u} = 0 & \text{in } \Omega, \\ \sigma(\mathbf{u}) \cdot \mathbf{n} = 0 & \text{on } \Sigma_1, \end{cases} \quad (1)$$

where $\mathbf{u}$ is the displacement, ρ the mass density, ω the pulsation, Ω the section of the guide and Σ_1 the inner and outer surfaces.

2.2. The SAFE method in curvilinear coordinates system

2.2.1. Overall principle. The Semi-analytical Finite Element (SAFE) method is considered as an efficient and well-adapted numerical method to compute guided waves in different guide geometries made of arbitrary material and section. This technique allows computing a series of guided modes by coupling finite element modeling technique in the guide section to an analytical approach to deal with the guided propagation. In other words, at a given frequency, the FE computation is used to solve the wave equation (1) in the cross section of the guide. Then, the dependency upon the propagation axis of the solution is handled analytically. The SAFE formulation leads to a quadratic eigenvalue system where the eigenvectors are the particle displacements and the eigenvalues are the wavenumbers of the calculated modes.

To use this approach, the following conditions must be satisfied:

- invariant cross section along the propagation axis ,
- invariant material properties,
- equilibrium equation independent of the propagator component.

Then, the displacement components of the elastodynamic field can be decomposed on eigenmodes calculated in the guide section by the SAFE method as follows:

$$\mathbf{u}(x,y,z) = \sum_n A_n \mathbf{u_n}(x,y,z) = \sum_n A_n \tilde{\mathbf{u}}_\mathbf{n}(x,y) e^{j(\omega t - \beta_n z)} \quad (2)$$

where ($\tilde{\mathbf{u}}_\mathbf{n}, \tilde{\boldsymbol{\sigma}}_\mathbf{n}, \beta_n, A_n$) is the displacement, stress, wave number and amplitude of each eigenmode.

In the case of a curved pipe, the SAFE approach must be derived in curvilinear coordinates in order to deal with the curvature effect. The present derivation is based on that proposed by Treyssède and Laguerre [2] who formulated the SAFE method to study GW propagation in a helical waveguide.

2.2.2. SAFE curvilinear formulation. In the elbow, the pipe centerline is defined by the position vector (see Fig. 2)

$$\mathbf{R}(s) = R_c\left[1-\cos(s/R_c)\right]\mathbf{e_x} + R_c\sin(s/R_c)\mathbf{e_z} \tag{3}$$

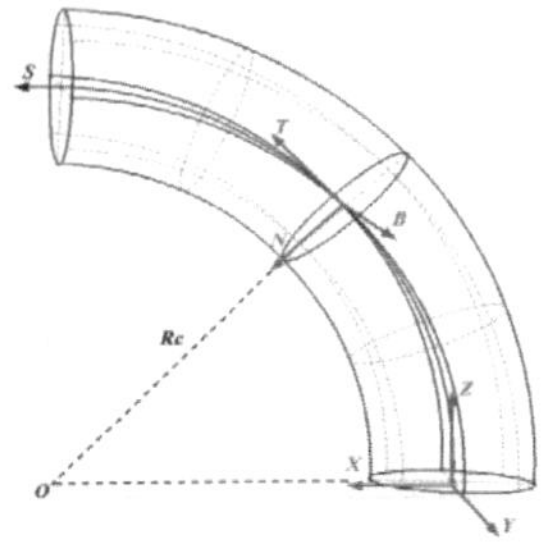

Figure 2. Schematic of an elbow with local curvilinear coordinates system

From this position vector, a new basis (**N**, **B**, **T**) is constructed, which represents the curvilinear basis associated to the configuration. The basis is defined by:

$$\mathbf{T} = \frac{d\mathbf{R}}{ds} \qquad \mathbf{N} = R_c\frac{d\mathbf{T}}{ds} \qquad \mathbf{B} = \mathbf{T}\wedge\mathbf{N} = \mathbf{e_y} \tag{4}$$

The non-orthonormal covariant basis associated to the curvilinear basis can then be defined as:

$$\mathbf{g}_1 = \mathbf{N} \qquad \mathbf{g}_2 = \mathbf{B} \qquad \mathbf{g}_3 = (1-\gamma x)\mathbf{T} \tag{5}$$

For this case, this covariant basis is orthogonal but the axes are no more normalized. The covariant metric tensor is given by:

$$\mathbf{g} = \begin{bmatrix} 1 & 0 & 0 \\ 0 & 1 & 0 \\ 0 & 0 & (1-\gamma x)^2 \end{bmatrix} \tag{6}$$

The variational formulation can be expressed in this covariant basis as follows:

$$\int_\Omega \boldsymbol{\delta\varepsilon}_{cov} : \boldsymbol{\sigma}_{cov} dV - \omega^2\int_\Omega \rho\boldsymbol{\delta}\mathbf{u}_{cov}\cdot\mathbf{u}_{cov} dV = 0 \tag{7}$$

Then, using the relation between the components of the elastodynamic field in the covariant and the curvilinear basis, respectively, each term of the previous formulation can be expressed in the (**N**, **B**, **T**) basis; the variational formulation might be re-written as

$$\int_\Omega \boldsymbol{\delta\varepsilon}^{\mathbf{T}}_{cur}\mathbf{C}\boldsymbol{\varepsilon}_{cur}\sqrt{\det(\mathbf{g})}\,dxdyds - \omega^2\int_\Omega \rho\boldsymbol{\delta}\mathbf{u}^{\mathbf{T}}_{cur}\mathbf{u}_{cur}\sqrt{\det(\mathbf{g})}\,dxdyds = 0 \tag{8}$$

with: $\mathbf{u}_{cur} = \mathbf{u}_{cur}(x,y)e^{i(ks-\omega t)}$ $\qquad \boldsymbol{\delta}\mathbf{u}_{cur} = \boldsymbol{\delta}\mathbf{u}_{cur}(x,y)e^{-i(ks-\omega t)}$.

In the equilibrium equation written in the curvilinear coordinate system, the s component does not explicitly appear. Hence, an exponential e^{iks} can be factorized in all elastodynamic field components. From this, we can write the following semi-analytical variational formulation by restricting the integration in the volume to a surface integration over the cross-section:

$$\begin{aligned}&\int_S \boldsymbol{\delta}\mathbf{u}^{\mathbf{T}}_{cur}\mathbf{L}^{\mathbf{T}}_{xy}\mathbf{C}\mathbf{L}_{xy}\mathbf{u}_{cur}\sqrt{\det(\mathbf{g})}dxdy + ik\int_S \boldsymbol{\delta}\mathbf{u}^{\mathbf{T}}_{cur}(\mathbf{L}^{\mathbf{T}}_{xy}\mathbf{C}\mathbf{L}_s - \mathbf{L}^{\mathbf{T}}_s\mathbf{C}\mathbf{L}_{xy})\mathbf{u}_{cur}\sqrt{\det(\mathbf{g})}dxdy \\ &+k^2\int_S \boldsymbol{\delta}\mathbf{u}^{\mathbf{T}}_{cur}\mathbf{L}^{\mathbf{T}}_s\mathbf{C}\mathbf{L}_s\mathbf{u}_{cur}\sqrt{\det(\mathbf{g})}dxdy - \omega^2\int_S \rho\boldsymbol{\delta}\mathbf{u}^{\mathbf{T}}_{cur}\mathbf{u}_{cur}\sqrt{\det(\mathbf{g})}dxdyds = 0\end{aligned} \tag{9}$$

where $\mathbf{L}_{xy}$ and $\mathbf{L}_s$ are two elementary matrices appearing in the strain-displacement relation:

$$\boldsymbol{\varepsilon} = (\mathbf{L}_{xy} + \frac{\partial}{\partial s}\mathbf{L}_s)\mathbf{u}$$

$$\mathbf{L}_{xy}=\begin{pmatrix}\frac{\partial}{\partial x} & 0 & 0\\ 0 & \frac{\partial}{\partial y} & 0\\ -\gamma(1-\gamma x) & 0 & 0\\ \frac{\partial}{\partial y} & \frac{\partial}{\partial x} & 0\\ 0 & 0 & 2\gamma+(1-\gamma x)\frac{\partial}{\partial x}\\ 0 & 0 & \frac{\partial}{\partial y}\end{pmatrix}\quad \mathbf{L}_{s}=\begin{pmatrix}0 & 0 & 0\\ 0 & 0 & 0\\ 0 & 0 & \frac{1}{(1-\gamma x)}\\ 0 & 0 & 0\\ \frac{1}{(1-\gamma x)} & 0 & 0\\ 0 & \frac{1}{(1-\gamma x)} & 0\end{pmatrix} \tag{10}$$

It must be noticed that the guide curvature appears in several coefficients of operators $\mathbf{L}_{\mathbf{xy}}$ and $\mathbf{L}_{\mathbf{s}}$. For a null curvature, these elementary matrices are equal to those of a straight guide.

Similarly to the SAFE method in a Cartesian basis, the numerical discretization (2D elements) of the waveguide cross section leads to the following quadratic system of equations to be solved:

$$(\mathbf{K}_1^e + j\beta\mathbf{K}_2^e + \beta^2\mathbf{K}_3^e)\mathbf{U} - \omega^2\mathbf{M}^e\mathbf{U} = 0 \tag{11}$$

$\mathbf{K}^e$ and $\mathbf{M}^e$ represent the elementary stiffness and mass matrices calculated at each element of the cross section of the guide. As shown previously, $\mathbf{U}$ and β correspond to the displacement vectors and the wavenumbers of each eigenmode.

2.3. Displacement field for a curved and straight pipe

To exemplify the effect of curvature, the three components of the displacement (radial, orthoradial and axial) of different propagative modes in an elbow are shown for different curvature values. The two modes chosen are the first torsional mode T(0,1) and the second extensional mode L(0,2) which are the modes the most commonly used in GW for non destructive inspections.

The elbows are 5mm-thick with an outer diameter of 10mm; the wave frequency is of 100 kHz. The curvature is varied from 0 (straight guide) to 1/15 mm^{-1}.

Fig. 3 shows the displacement field for the T(0,1) mode.

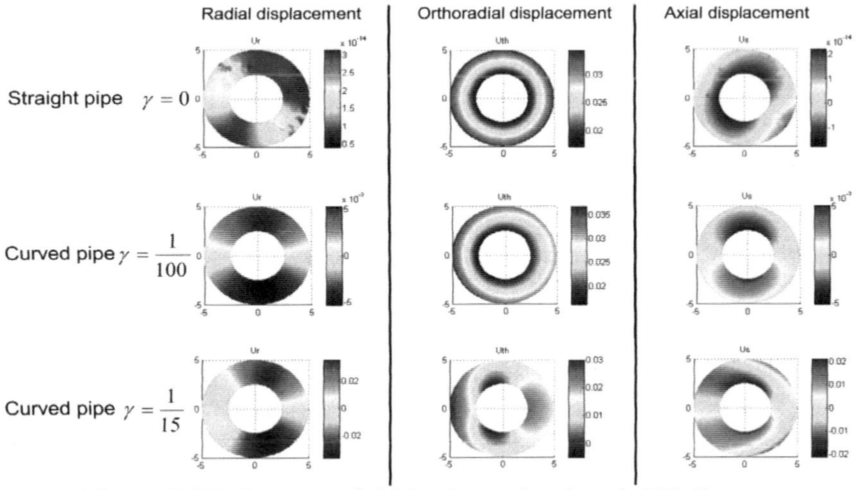

Figure 3. Displacement field for the torsional mode T(0,1)

A non-zero curvature causes distortions of the displacement field that increases for increasing curvature. The axisymmetry of orthoradial displacement the mode in the straight pipe is lost in a curved guide. This component is the only one carrying energy in the straight pipe (the maps for other components are shown at an amplitude scale 10^{-12} lower than that of the orthoradial component, corresponding to numerical errors). In a curved pipe, the two other components of the displacement increase with increasing curvature and are eventually of the same order of magnitude as that of the orthoradial component for the highest curvature. Fig. 4 shows similar results for the L(0,2) mode.

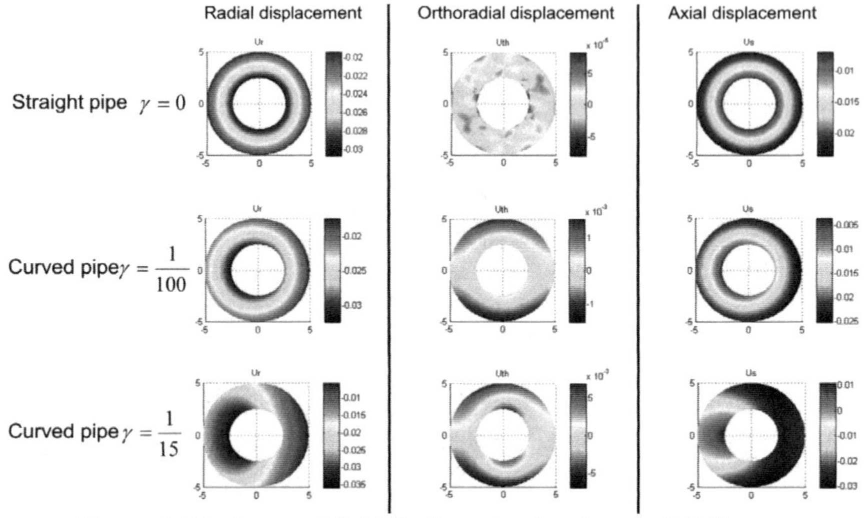

Figure 4. Displacement field for the extensional mode L(0,2)

For this extensional mode, the radial and axial displacements fields dominate in the straight guide. Distortions are observable again as soon as a non-zero curvature is considered.

Similar effects of non-zero curvature can be observed when studying flexural modes.

In addition, for a straight pipe, the flexural modes appear as pairs of modes having the same wavenumber but with displacement components rotated by an angle of $\pi/2$. Conversely, for a curved pipe, these flexural modes are separated, because of the non-axisymmetry of the guide (see fig. 5). It is also observed that the cut-off frequency of each propagative mode in a curved case increases in comparison with the straight case. In Fig. 5, dispersion curves of GW in a straight steel pipe are superimposed with dispersion curves of GW in an elbow.

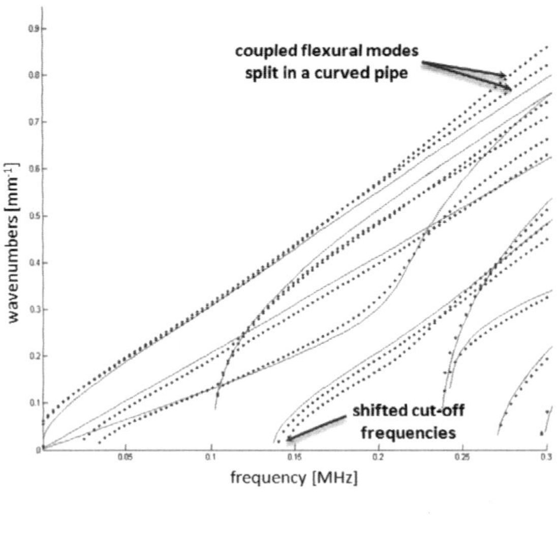

Figure 5. Wavenumbers dispersion curves for a straight (continuous line) and a curved steel pipe (Outer diameter=10mm, Thickness=2.5mm) of 1/25 mm^{-1} curvature (dotted line)

3. The mode matching model at the common junction between two guides

3.1. Analytical mode matching formulation:

In pipeworks, reflection and transmission phenomena arise at each junction of two different parts of the pipe. In the case of an elbow, the scattering phenomena arise at the junction of a straight and a curved part of the guide.

The mode matching technique allows computing the modal decomposition of an incident mode onto reflected and transmitted ones. An interesting mode matching formulation for the computation of Lamb wave scattering matrices has been investigated by Feng *et al.* [3]. It is based on the use of mixed stress-displacement vectors **X** and **Y** introduced by Pagneux and Morel [4] and defined as follows:

$$\mathbf{X} = \begin{pmatrix} \mathbf{t}_S \\ u_z \end{pmatrix} = \begin{pmatrix} \sigma_{xz} \\ \sigma_{yz} \\ u_z \end{pmatrix}, \quad \mathbf{Y} = \begin{pmatrix} \mathbf{u}_S \\ t_z \end{pmatrix} = \begin{pmatrix} u_x \\ u_y \\ -\sigma_{zz} \end{pmatrix}. \tag{12}$$

In our case, the mode matching model will be computed at the junction of a straight pipe and a curved one. However, the formal solution is generic and can be easily applied to other cases of junction (curved-curved, straight-straight).

The mode-matching method consists basically in satisfying the continuity between the elastodynamic fields associated to the straight and the curved parts of the guide. Mathematically, it involves the calculation of scalar product coming from the biorthogonality relation linking the displacement and stress components of each elastodynamic fields considered. These scalar products are easily calculated using the bi-orthogonality relation between the hybrid vectors.

Therefore, using the eigenmodes obtained by the SAFE model in the straight and the curved parts, GW displacement and stress fields can be expressed as modal series. From now on, hybrid vectors are explicitly used. The total fields in the straight part (1) and in the curved one (2) can be expressed as the superposition of the calculated eigenmodes:

$$\begin{cases} \begin{pmatrix} \mathbf{X}_{tot1} \\ \mathbf{Y}_{tot1} \end{pmatrix} = \sum_n B_n^1 \begin{pmatrix} \mathbf{X}_{n1}^- \\ \mathbf{Y}_{n1}^- \end{pmatrix} \mathrm{e}^{-\mathrm{j}\beta_{n1}^- z} + \sum_n A_n^1 \begin{pmatrix} \mathbf{X}_{n1}^+ \\ \mathbf{Y}_{n1}^+ \end{pmatrix} \mathrm{e}^{-\mathrm{j}\beta_{n1}^+ z} \\ \begin{pmatrix} \mathbf{X}_{tot2} \\ \mathbf{Y}_{tot2} \end{pmatrix} = \sum_n A_n^2 \begin{pmatrix} \mathbf{X}_{n2}^+ \\ \mathbf{Y}_{n2}^+ \end{pmatrix} \mathrm{e}^{-\mathrm{j}\beta_{n2}^+ s} + \sum_n B_n^2 \begin{pmatrix} \mathbf{X}_{n2}^- \\ \mathbf{Y}_{n2}^- \end{pmatrix} \mathrm{e}^{-\mathrm{j}\beta_{n2}^- s} \end{cases} \tag{13}$$

where (A_n^1, B_n^1) and (A_n^2, B_n^2) are the amplitude of the incoming and out-coming modes calculated in the straight and the curved parts of the pipe (as shown by Fig. 5)

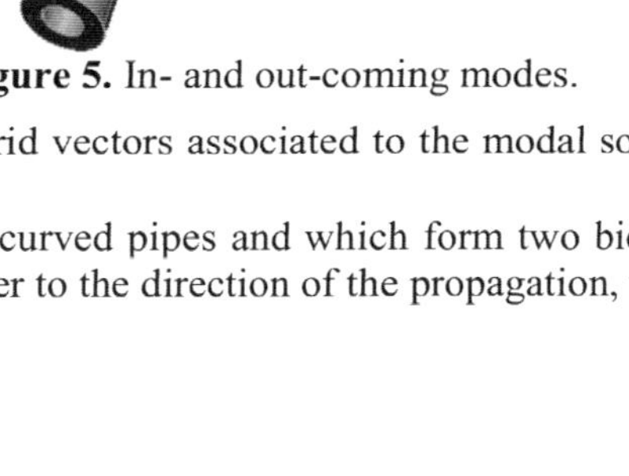

Figure 5. In- and out-coming modes.

$\begin{pmatrix} \mathbf{X}_{n1} \\ \mathbf{Y}_{n1} \end{pmatrix}$ and $\begin{pmatrix} \mathbf{X}_{n2} \\ \mathbf{Y}_{n2} \end{pmatrix}$ are the hybrid vectors associated to the modal solutions obtained by the SAFE calculation in the straight and the curved pipes and which form two biorthogonal bases. The index – (out-coming) and + (incoming) refer to the direction of the propagation, with:

$$\mathbf{X}_{n1}^{-} = -\mathbf{X}_{n1}^{+} \quad \text{and} \quad \mathbf{Y}_{n1}^{-} = \mathbf{Y}_{n1}^{+}$$
$$\mathbf{X}_{n2}^{-} = -\mathbf{X}_{n2}^{+} \quad \text{and} \quad \mathbf{Y}_{n2}^{-} = \mathbf{Y}_{n2}^{+}$$
$$\beta_{n1}^{-} = -\beta_{n1}^{+} \quad \text{and} \quad \beta_{n2}^{-} = -\beta_{n2}^{+}$$

In the mode matching formulation, $(\cdot|\cdot)_S$ denotes the following scalar product:

$$\left(\mathbf{Y}|\,\mathbf{X}\right)_S = \int_S (\mathbf{t}_S \cdot \mathbf{u}_S + t_z u_z)\, dS \text{ ,(14)} \tag{14}$$

where S is the cross-section of the guide.

Then, the continuity condition of elastodynamic fields must be satisfied as follows:

$$\begin{pmatrix} \mathbf{X}_{tot1} \\ \mathbf{Y}_{tot1} \end{pmatrix} = \begin{pmatrix} \mathbf{X}_{tot2} \\ \mathbf{Y}_{tot2} \end{pmatrix} \quad \text{at } z = s = z_0, \tag{15}$$

re-written as:

$$\sum_n B_n^1 \begin{pmatrix} -\mathbf{X}_{n1} \\ \mathbf{Y}_{n1} \end{pmatrix} e^{-j\beta_{n1}^{-} z} + \sum_n A_n^1 \begin{pmatrix} \mathbf{X}_{n1} \\ \mathbf{Y}_{n1} \end{pmatrix} e^{-j\beta_{n1}^{+} z} = \sum_n A_n^2 \begin{pmatrix} \mathbf{X}_{n2} \\ \mathbf{Y}_{n2} \end{pmatrix} e^{-j\beta_{n2}^{+} s} + \sum_n B_n^2 \begin{pmatrix} -\mathbf{X}_{n2} \\ \mathbf{Y}_{n2} \end{pmatrix} e^{-j\beta_{n2}^{-} s} \tag{16}$$

Finally, the mode matching will consist in the computation of the scattering matrix **S** that links the amplitudes of the incoming modes to those of the out-coming modes at the shared junction:

$$\begin{pmatrix} \mathbf{B}_1 \\ \mathbf{A}_2 \end{pmatrix} = \mathbf{S} \begin{pmatrix} \mathbf{A}_1 \\ \mathbf{B}_2 \end{pmatrix} \tag{17}$$

where

$$\mathbf{S} = \begin{pmatrix} \mathbf{R}_{1\to2} & \mathbf{T}_{2\to1} \\ \mathbf{T}_{1\to2} & \mathbf{R}_{2\to1} \end{pmatrix} \tag{18}$$

$\mathbf{R}_{1\to2}, \mathbf{T}_{1\to2}$ ($\mathbf{R}_{2\to1}, \mathbf{T}_{2\to1}$, respectively) are the reflection and transmission matrices of a wave arising from the straight part (from the curved part, respectively).

3.2. Scattering matrix calculation

According to Feng *et al.* [3], the scattering matrix can be defined as

$$\mathbf{S} = \begin{pmatrix} 2\mathbf{J}_1 & \mathbf{J}_{21}^T - \mathbf{J}_{12} \\ \mathbf{J}_{12}^T - \mathbf{J}_{21} & 2\mathbf{J}_2 \end{pmatrix}^{-1} \begin{pmatrix} [0]_{NxN} & \mathbf{J}_{12} + \mathbf{J}_{21}^T \\ \mathbf{J}_{12}^T + \mathbf{J}_{21} & [0]_{NxN} \end{pmatrix} \tag{19}$$

N denotes the number of modes used for the mode matching, including all the propagative modes at the working frequency and a number of evanescent modes. $\mathbf{J}_1$ and $\mathbf{J}_2$ are projection matrices obtained from the biorthogonality relation used for each field separately and $\mathbf{J}_{12}$ and $\mathbf{J}_{21}$ represent projection matrices which couple the stress and displacement components of both elastodynamic fields.

In the present case, the mode matching formulation must handle fields in both straight and curved guides which are described in two different coordinate systems. Each field involves a specific scalar product adapted to each geometry. The key point of our mode matching model is to compute the appropriate projection matrices $\mathbf{J}_{12}$ and $\mathbf{J}_{21}$ dealing with both two different bases.

$\mathbf{J}_{12}$ *and* $\mathbf{J}_{21}$ *computation*

Assuming the completeness of the eigenmodes obtained by the SAFE calculation, the biorthogonality relation (Fraser's relation) allows expressing the biorthogonality between two propagative modes m and n in the same guide through the bi-linear relation given as

$$(\mathbf{X}_n|\mathbf{Y}_m)_S = \int_S (\mathbf{t}_S \mathbf{u}_S + t_z u_z)\, dx\, dy = \mathbf{J}_n \delta_{nm} \tag{20}$$

From this relation, the projection matrices $\mathbf{J}_1$ and $\mathbf{J}_2$ of both the straight and the curved parts of the guide can be numerically computed and we get:

$$(\mathbf{X}_n|\mathbf{Y}_m)_S = \hat{\mathbf{X}}_n^{\mathbf{T}} \mathbf{M}_S \hat{\mathbf{Y}}_m = \mathbf{J}_n \delta_{nm} \tag{21}$$

$\mathbf{M}_S$ is the projection matrix obtained by combining the interpolation matrices $N_e(x,y)$ used to compute the elastodynamic field at each element of the mesh, $\hat{\mathbf{X}}_n^{\mathbf{T}}$ and $\hat{\mathbf{Y}}_m$ being the nodal hybrid vectors.

Since two different coordinate systems are used (Cartesian and curvilinear) to compute the modal solutions in the two different parts of the pipe involving two different scalar products associated to each basis.

In order to use the mode matching approach presented here, we need to calculate mixed scalar products by combining hybrid vectors stemming from the different elastodynamic fields. Hence, the projection of a hybrid vector of a mode n of the first elastodynamic field denoted $\mathbf{X}_n^1(x,y,s)$ on the hybrid vector of a mode m denoted $\mathbf{Y}_m^2(x,y,z)$ of the second one can be expressed as follows:

$$(\mathbf{X}_n^1 \,|\, \mathbf{Y}_m^2)_S = \int_{V_c} \mathbf{X}_n^1(x,y,s)\mathbf{Y}_m^2(x,y,s)\,dV_1 \tag{22}$$

We can observe that the elementary volume of integration dV_1 is associated to the projected component. For the mode matching method, this volume must be expressed in the orthonomal basis. In the present case, because of the use of two coordinate systems, we must express this elementary volume in both the curvilinear and the Cartesian bases.

For a curved guide, the elementary volume corresponding to the coordinate differentials (dx, dy, ds) is given by:

$$dV_1 = |\mathbf{g}_1.(\mathbf{g}_2 \wedge \mathbf{g}_3)|\,dxdyds = \sqrt{\det(\mathbf{g})}dxdyds \tag{23}$$

with the symbol $\wedge$ denoting the vector cross product and where $(\mathbf{g}_1, \mathbf{g}_2, \mathbf{g}_3)$ are the vectors of the covariant basis associated to the curvilinear coordinate system previously showed in section 2.

Interestingly, the straight case is a specific case of the curved one where $\sqrt{\det(\mathbf{g})} = 1$, then, the elementary volume definition still the same in a Cartesian coordinate system.

Hence from (22) and (23), considering the hybrid vectors of an eigenmode n of the elbowed pipe $\mathbf{X}_n^c(x,y,s)$ and an eigenmode m $\hat{\mathbf{Y}}_m^s(x,y)$ of the straight guide, the scalar product is defined as:

$$(\mathbf{X}_n^c \,|\, \mathbf{Y}_m^s)_S = \int_S \mathbf{X}_n^c(x,y,s)\mathbf{Y}_m^s(x,y)\sqrt{\det(\mathbf{g})}dxdyds = \mathbf{X}_n^{c\mathbf{T}}\mathbf{M}_s^c\mathbf{Y}_m^s = \mathbf{J}_{21}\delta_{nm} \tag{25}$$

$\mathbf{M}_s^c$ corresponds to the global mass matrix (without the constant ρ) used in the SAFE formulation in the curvilinear coordinate system.

Similarly, the projection of an elastodynamic hybrid vector in the straight part $\mathbf{X}_n^s(x,y,s)$ in an eigenmode in the curved one $\hat{\mathbf{Y}}_m^c(x,y)$ can be expressed as:

$$(\mathbf{X}_n^s \,|\, \mathbf{Y}_m^c)_S = \int_S \mathbf{X}_n^s(x,y,s)\,\mathbf{Y}_m^c(x,y)\sqrt{\det(\mathbf{g})}\,dxdyds = \mathbf{X}_n^{s\mathbf{T}}\mathbf{M}_s^s\mathbf{Y}_m^c = \mathbf{J}_{12}\delta_{nm} \tag{26}$$

Now, it is possible to calculate $\mathbf{J}_1$ and $\mathbf{J}_2$ using (21) and $\mathbf{J}_{12}$and $\mathbf{J}_{21}$ using (25). Therefore, the scattering matrix **S** presented in (18) can be computed.

4. Numerical results

4.1. Configuration

The straight-curved-straight pipe shown on Fig. 6 and describe in the figure caption is studied.

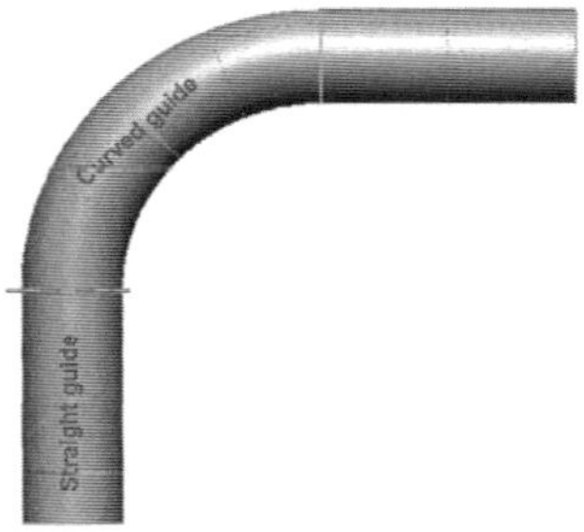

Figure 2. Two straight steel pipes of outer diameter equal to 10 mm and 2.5 mm-thick are linked by a curved pipe of 1/15 mm^{-1} curvature. The examination is made at a frequency of 100 kHz.

4.2. SAFE calculations
The eigenmodes are calculated using the appropriate SAFE model for each part of the guide. At the frequency considered the cut-off frequency of each propagative mode in the curved case increases as compared to the straight case. Therefore, the number of propagative modes in the straight guide and the curved one may differ.

SAFE calculation in the straight guides: at 100 kHz, six propagative modes in the straight parts are predicted: 4 flexural modes F(1,1), F(-1,1), F(1,2), and F(-1,2), 1 torsional mode T(0,1), 1 extensional mode: L(0,1). The flexural modes F(1,1) and F(-1,1) (respectively F(1,2) and F(-1,2)) are calculated in pairs and represent the same flexural mode in the straight case.

SAFE calculation in the curved guide: at 100 kHz, four propagative modes in the curved part are predicted: 2 flexural-like modes: F(1,1) and F(-1,1), 1 torsional-like mode: T(0,1), 1 extensional-like mode: L(0,1)

4.3. Modal decomposition
Scattering phenomena arise at the junctions when propagative modes coming from the straight part reach the curved one. The incident modes give rise to reflected and transmitted modes.

Considering all possible propagative modes coming from the straight guide, the energy partition of the incident energy into transmitted modes in the curved part is detailed in the following table:

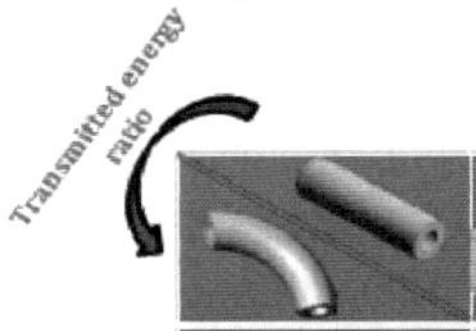

	F(1,1)	F(-1,1)	T(0,1)	L(0,1)	F(1,2)	F(-1,2)
F(1,1)	54,14%	17,65%	23,95%	2,89E-06%	1,00%	2,04%
F(-1,1)	20,98%	64,29%	1,23E-06%	10,80%	2,43%	1,17%
T(0,1)	15,59%	5,08%	73,25%	7,20E-06%	1,55%	3,18%
L(0,1)	2,92%	8,97%	6,33E-06%	82,61%	1,55%	0,74%

Figure 7: Energy partition of transmitted modes in the curved part for the various possible incident modes in the straight part.

This table illustrates the ability of the mode matching model to calculate the modal decomposition of an arbitrary incident mode at the junction of two guides by predicting its modal decomposition. Here, each propagative mode is decomposed onto the modal basis of the curved guide and the transmission is higher into modes in the curved which are of the same nature as that of the incident mode in the straight guide. However, a torsional or extensional incident mode (as typically used in NDT operations) partially converts into transmitted flexural eigenmodes. The modal decomposition of the reflected energy can be obtained using the same model.

In addition, from the same scattering matrix, it is possible to extract the modal decomposition of an incident mode arising from the curved guide into reflected and transmitted modes.

4.4. Curvature effect
In this section, the effect of the curvature is investigated. Here, different curvatures are considered and the mode matching model is used to predict the transmitted and reflected energy ratios for an incoming mode. In the following example, the incident mode is the first extensional L(0,1) in the straight guide.

The total energy (red line on Fig. 8) is conserved, even in the case of the 1/6 mm^{-1} curvature which leads to a radius of the intrados as low as 1 mm (to be compared to the outer diameter of 10 mm). The reflected energy increases for increasing elbow curvature.

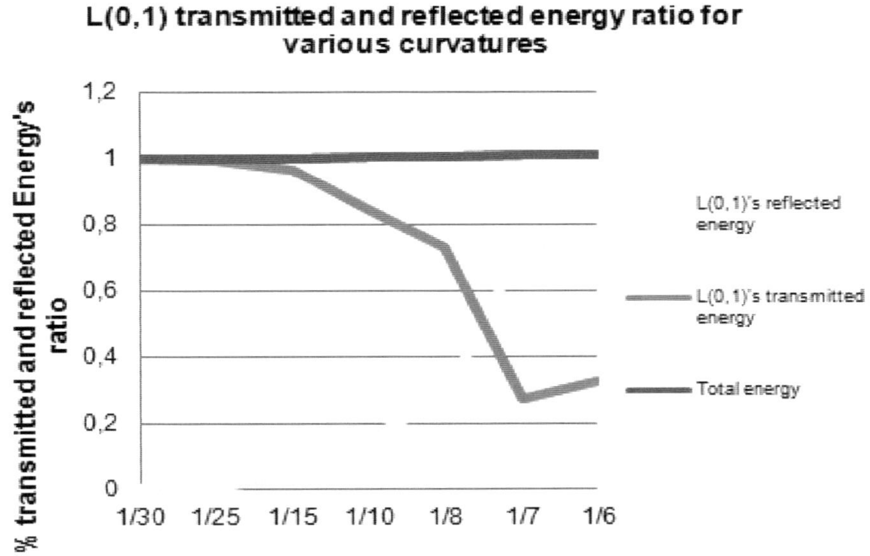

Figure 8. Reflected and transmitted energy ratio at a straight-curved junction for a L(0,1) incident mode in a steel pipe (OD=10mm, Th=2.5mm) as functions of the curvature of the elbow at 100kHz.

Similarly, for the other propagative modes (torsional and flexurals), the transmitted and reflected energies variations are represented as follows:

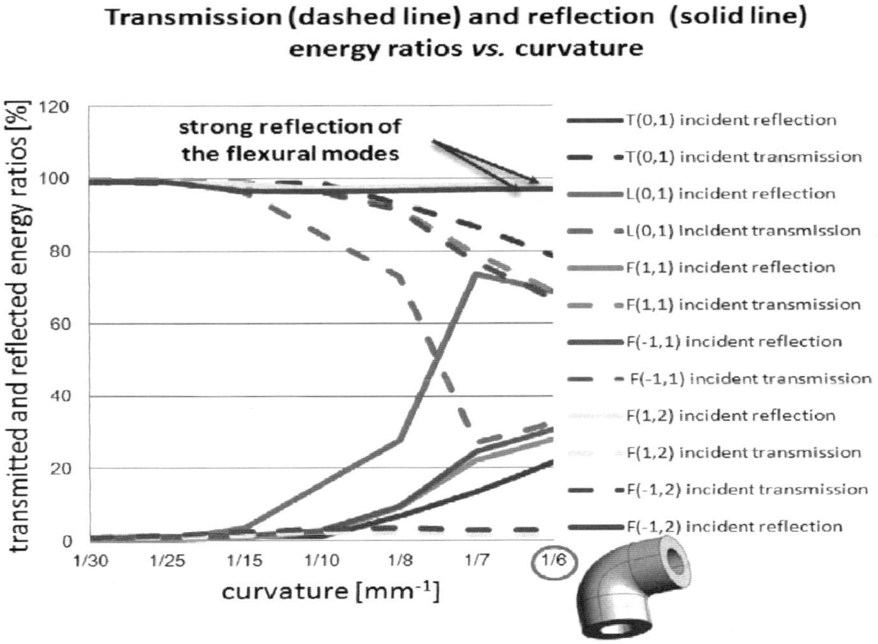

Figure 9. Reflected and transmitted energy ratios at a straight-curved junction for all possible incident modes in a steel pipe (OD=10mm, Th=2.5mm) as functions of the curvature of the elbow at 100kHz.

Globally, the evolution of reflected and transmitted energy ratios has a similar behavior compared to the L(0,1) case, except for the flexural modes F(1,2) and F(-1,2) which are strongly reflected, whatever the curvature. The reason for this behavior is that, at the frequency considered for the examples, these two modes have no equivalent into the curved guide, due to the fact that the cut-off frequencies of these modes in the curved part are higher than those in the straight part and the working frequency is in between cut-off frequency in the straight part and those in the curved part.

Thus, the choice of the frequency used for inspecting a pipework containing curved parts must be considered as an important criterion in the mode transmission through the pipe elbow.

5. Conclusions

In this study, guided waves propagation in a pipework with curved part has been investigated. In a first step, the SAFE method has been derived to compute guided wave propagation in a curved guide. Adapting the formulation proposed in [2] to the case of a curved pipe, eigenmodes expressed in a curvilinear coordinate system have been studied. Parametric studies have shown that the curvilinear SAFE implementation made can accurately deal with arbitrarily curved parts.

Then, the scattering phenomena at the junction of a straight guide and a curved one have been considered. Thanks to the mode matching method, the scattering matrix, that contains transmission and reflection coefficients and links two elastodynamic fields computed in different coordinate systems, has been computed given as modal series. The mode matching model formulation derived herein is generic and can deal with various coordinate systems by using adequate change of basis in regard to the configurations studied. Finally, parametric studies have been presented to illustrate the mode matching model capabilities.

Future work will include the calculation of the global scattering matrix resulting from the combination of several scattering matrices (welded elbow) according to [5], as well as experimental validations.

6. References

[1] Hayashi T, Kawashima K, Sun Z and Rose J L 2005 Guided wave propagation mechanics across a pipe elbow *J. Press. Vessel Tech.* **127** 322

[2] Treyssède F and Laguerre L 2010 Investigation of elastic modes propagating in multi-wire helical waveguides *J. Sound Vib* **329** 1702

[3] Feng F, Shen J and Lin S 2012 Scattering matrices of lamb waves at irregular surface and void defects *Ultrasonics* **52** 760

[4] Pagneux V and Maurel A 2006 Lamb wave propagation in elastic waveguides with variable thickness *Proc. R. Soc.* **A 462** 1315

[5] Baronian V, Lhémery A and Jezzine K 2011 Hybrid SAFE/FE simulation of inspections of elastic waveguides containing several local discontinuities or defects *Rev. Prog. QNDE* **30** 183

EMAT generation of bulk forces in a ferromagnetic plate and their equivalent surface stresses *

C Rouge[1], A Lhémery[1] and C Aristégui[2]

[1] CEA, LIST, F-91191 Gif-sur-Yvette *cedex*, France
[2] Université de Bordeaux, UMR CNRS 5295, I2M/Acoustique Physique, F-33405 Talence *cedex*, France

E-mail: alain.lhemery@cea.fr

Abstract. Electro-magnetic acoustic transducers (EMAT) are successfully used in many NDE applications, despite their low efficiency: they do not require a coupling medium and can easily generate elastic waves that standard piezoelectric transducers cannot, such as shear horizontal guided waves. There are all sorts of EMAT designs, so much so that dedicated simulation tools are necessary to optimally conceive an EMAT for a given application. EMAT performances also strongly depend on material properties of the piece under test. Here, ferromagnetic materials are considered. In such a material, an EMAT is the source of three forces resulting from three distinct and generally nonlinear phenomena: in addition to the Lorentz's force generated in all conductive media, the magnetization and magnetostriction forces take place. All these forces are modelled as vector fields in the volume of the specimen. However, wave generation is more efficiently predicted by considering sources of surface stress than sources of body force. Thus, a general model is derived for transforming body forces into surface stresses; this approach is used to express the 2D modal amplitudes of Lamb waves generated by an EMAT in a ferromagnetic plate as quasi-closed form solutions.

1. Introduction

Electro-magnetic acoustic transducers (EMAT) are a common way to generate and receive guided waves, especially shear horizontal waves in plate and torsional waves in cylinders which are not easily generated by piezoelectric transducers. Counterbalancing both their low signal-to-noise ratio and the fact that they only work on conductive media, EMAT allow high speed inspections since they do not require a coupling medium. They can also be beneficially used in hostile environments (high temperature and/or high pressure), in which piezoelectric transducers fail.

The versatility of EMAT design is very attractive too. By choosing adequately the excitation frequency and the arrangement of the electric coil relatively to the magnetization direction provided by one (or several) permanent magnet(s), one mode and its associated polarization among the various propagative guided modes can be selected [1]. Simulation tools are therefore helpful to optimize the various parameters in the EMAT design.

In conductive media, EMAT generate ultrasonic waves through the well known Lorentz's force. In ferromagnetic media, two supplementary forces are created and must be taken into account when modelling the wave generation: these are the magnetization and magnetostriction forces. These forces

* This paper corresponds to a conference given during AFPAC 2012 and should have been published in *J. Phys.: Conf. Ser.* **457** with other papers of the same year of the series.

result from the complex behaviour of ferromagnetic material under magnetization and may constitute the predominant phenomena of wave generation mechanism. Considering a ferromagnetic medium implies to take into account magnetic non linear behaviours such as hysteresis, saturation and double frequency effects. Many phenomena occur, which possibly interact between each other. For example, magnetostriction couples two effects: when a magnetic field is applied to a ferromagnetic material, the medium changes its physical dimensions; at the same time, this change in dimensions modifies its magnetic field.

All these three electromagnetic forces, Lorentz, magnetization and magnetostriction, appear as sources of body forces in the equation of wave motion. Therefore, solving the equation requires performing volume integrations to predict the waves they generate. To avoid time-consuming calculations and to reuse previously developed models of guided wave generation by a distribution of surface stresses [2], body forces are thus transformed into surface stresses. The transformation follows a method described by Thompson [3]; it also ensures the mixing of electromagnetic phenomena (generated forces) with wave generation and propagation phenomena through the Green's integral formulation of the elastic wave equation.

The EMAT is modelled as being made of a permanent magnet that creates a static magnetic induction field, and of an electric coil that creates a dynamic magnetic induction field. In a first part, from the static and dynamic induction fields, the excitation field and the magnetization field are expressed. The three electromagnetic forces are then written analytically. A discussion about the modelling of the double frequency effect, the saturation and the hysteresis phenomena is also made. These forces are then transformed into surface stresses. As an application, the modal amplitudes of Lamb waves are obtained as quasi-closed form solutions in the 2D case. Examples of calculation are then given and discussed. In what follows, developments are written in the Cartesian coordinate system because only guided waves generated in a plate (Lamb waves) will be considered.

2. Theory

We develop a semi-analytical model for simulating the generation of elastic waves in a ferromagnetic material by an EMAT. The magnetic fields generated by the EMAT and the resulting forces are first derived; then, equivalent surface stresses are obtained through a transformation that generalizes a transformation already used in [4] for SH guided waves.

2.1. Magnetic field created by an EMAT

An EMAT is made of a permanent magnet combined with an electrical coil: the first component creates a static magnetic induction field in the material while the second creates a dynamic one. These fields are denoted by $\vec{B}_S$ and $\vec{B}_D$, respectively. The dynamic field has the same frequency content as that of the current injected in the electric coil. From now on, a continuous wave excitation $e^{i\omega t}$ is considered, where ω denotes the angular frequency [rad.s^{-1}].

The other two magnetic fields, the excitation field $\vec{H}$ and the magnetization field $\vec{M}$, two necessary quantities to predict the electromagnetic forces created by an EMAT in a ferromagnetic material, are derived from the induction field $\vec{B}$ through the two following relations:

$$\vec{B} = \mu_0 \bar{\bar{\mu}} \vec{H} \ , \tag{1}$$

$$\vec{M} = \chi \vec{H} \ , \tag{2}$$

where $\mu_0 = 4\pi 10^{-7}$H.m^{-1} is the vacuum permeability. The magnetic permeability $\bar{\bar{\mu}}$ is a diagonal second rank tensor while the magnetic susceptibility χ is a scalar constant.

The magnetic permeability of a ferromagnetic material depends both on the amplitude of the applied magnetic induction [5] and on its frequency [6]. In what follows, the permeability is given by two different values. If the dynamic and static fields have a non zero component in the same direction, the static value of the permeability is used because the static induction field is generally much higher than the dynamic one, in standard applications of EMAT. For example, if the wires of the electric coil are along the $\vec{y}$ direction, the dynamic field is along the $\vec{x}$ and $\vec{z}$ directions. If a static field along the $\vec{z}$ direction is superimposed, the permeability used in the $\vec{x}$ direction is the dynamic one and that used in the $\vec{z}$ direction is the static one. Assuming that a model for the dependencies of $\bar{\bar{\mu}}$ on the amplitude and on the frequency of the induction field exists, it could be introduced in our overall approach at no additional cost.

2.2. Body forces generated by an EMAT into a ferromagnetic medium
The forces generated by an EMAT in a ferromagnetic material can now be computed as they result from the magnetic fields described in the previous paragraph.

2.2.1. *Lorentz's force.* This force is generated in all conductive media. It results from the interaction of eddy currents created by the electric coil of the EMAT and the induction field $\vec{B}$ created by both the permanent magnet and the electric coil. It is expressed by:

$$\vec{F_L} = \vec{J} \wedge \vec{B} \ , \tag{3}$$

where $\vec{J}$ is the induced current density [A.m^{-2}].

2.2.2. *Magnetization force.* The magnetization force:

$$\vec{F_M} = \vec{M}.\bar{\bar{\nabla}}\vec{H} \tag{4}$$

directly results from the Bloch wall motion and the magnetic moments rotation of a ferromagnetic material under magnetization [5].

2.2.3. Magnetostriction force. This force results from the magnetization of a ferromagnetic material. If a magnetic field is applied to a ferromagnetic medium, the medium geometry changes; the reciprocal effect has been shown to make a negligible contribution in EMAT transduction processes [7]. Magnetostriction combines magnetic and elastic phenomena; it is commonly assumed that it does not induce an overall volume change.

The magnetostriction is modelled from the theory proposed by Hirao and Ogi [8]. The associated force:

$$\overrightarrow{F_{MS}} = \vec{\nabla}.\bar{\bar{\sigma}}^{MS} \tag{5}$$

depends on the magnetostriction stress $\bar{\bar{\sigma}}^{MS}$,

$$\sigma_{ij}^{MS} = -e_{ijm}H_m \ , \tag{6}$$

which is defined from the piezomagnetic stress tensor $\bar{\bar{e}}$,

$$e_{ijm} = C_{ijkl}d_{klm} \ , \tag{7}$$

where $\bar{\bar{C}}$ is the stiffness tensor and $\bar{\bar{d}}$ is the piezomagnetic strain coefficient tensor given by:

$$d_{klm} = \frac{\partial \varepsilon_{kl}}{\partial H_m} \ . \tag{8}$$

$\bar{\bar{\varepsilon}}$ denotes the magnetostrictive strain tensor. In this model, the values of the last tensor may be deduced from experimental data. For example, Hirao and Ogi [8] use only one measured curve of ε as a function of *H*, assuming that no magnetic anisotropy occurs, so that a magnetic microscopic description is not necessary. Their magnetostriction curve corresponds to the deformation along the total field. Then, they use a rotation of the strain tensor in order to express the magnetostrictive strain tensor $\bar{\bar{\varepsilon}}$ as a function of the interpolated experimental magnetostriction curve and its derivative, in a two dimensions configuration.

2.3. Magnetic non linear effect modelling: double frequency and saturation effects

2.3.1. Double frequency effect. This effect is generally negligible in classical EMAT applications and not modelled. Experiments [9] confirmed that the doubling frequency effect is negligible when the static field in the material is much larger than the dynamic one. This can be easily explained as follows.

First, considering the static and dynamic inductions, equations (1) and (2) yield:

$$\overrightarrow{B_S} = \mu_0 \bar{\bar{\mu}} \overrightarrow{H_S} \ , \tag{9}$$

$$\overrightarrow{B_D} e^{i\omega t} = \mu_0 \bar{\bar{\mu}} \overrightarrow{H_D} e^{i\omega t} \ , \tag{10}$$

$$\overrightarrow{M_S} = \chi \overrightarrow{H_S} \ , \tag{11}$$

$$\overrightarrow{M_D} e^{i\omega t} = \chi \overrightarrow{H_D} e^{i\omega t} \ , \tag{12}$$

where the induction field is simply decomposed as the sum:

$$\vec{B} = \overrightarrow{B_S} + \overrightarrow{B_D} e^{i\omega t} \ . \tag{13}$$

The Lorentz's force is then expressed by:

$$\overrightarrow{F_L} = \vec{J} e^{i\omega t} \wedge \left(\overrightarrow{B_S} + \overrightarrow{B_D} e^{i\omega t}\right) = \vec{J} \wedge \overrightarrow{B_S} e^{i\omega t} + \vec{J} \wedge \overrightarrow{B_D} e^{i2\omega t} \ . \tag{14}$$

The current density is a pure dynamic quantity resulting from the dynamic current injected in the electric coil. In this expression, it is obvious that if the static field dominates the dynamic field, the double frequency effect is negligible.

For the magnetization force, the doubling frequency is a bit more hidden in the equations. We have:

$$\begin{aligned} \overrightarrow{F_M} &= \mu_0 \left(\overrightarrow{M_S} + \overrightarrow{M_D} e^{i\omega t}\right) \cdot \left(\overline{\overline{\nabla H_S}} + \overline{\overline{\nabla H_D}} e^{i\omega t}\right) \\ &= \mu_0 \overrightarrow{M_S} \cdot \overline{\overline{\nabla H_S}} + \mu_0 \left(\overrightarrow{M_S} \cdot \overline{\overline{\nabla H_D}} + \overrightarrow{M_D} \cdot \overline{\overline{\nabla H_S}}\right) e^{i\omega t} + \mu_0 \overrightarrow{M_D} \cdot \overline{\overline{\nabla H_D}} e^{i2\omega t} \ . \end{aligned} \tag{15}$$

Here, the gradient of the excitation field is involved. If the static field can be considered as being uniform, as it is the case in many EMAT designs, the magnetization force reduces to:

$$\overrightarrow{F_M} = \mu_0 \overrightarrow{M_S} \cdot \overline{\overline{\nabla H_D}} e^{i\omega t} + \mu_0 \overrightarrow{M_D} \cdot \overline{\overline{\nabla H_D}} e^{i2\omega t} \ , \tag{16}$$

and again, the double frequency effect can be neglected if the amplitude of the static field is much higher than that of the dynamic one. Under this assumption, the case of periodic permanent magnet EMAT cannot be treated.

The same analysis cannot be made for the magnetostriction force because the analytical expression of the magnetostriction curve was derived from the approximation of an experimental curve. This curve depends on static and dynamic field amplitudes so that the piezomagnetic stress coefficients matrix $\bar{\bar{e}}$ also depends on both these magnetic fields. The frequency-dependent behaviour is therefore complex for low amplitude of the static field as $\bar{\bar{\varepsilon}}$ shows strong variations in this domain. At higher static amplitude, $\bar{\bar{\varepsilon}}$ varies almost linearly so that its derivative is constant; as a consequence, the frequency-dependency of the force becomes proportional to that of the induction and varies as $e^{i\omega t}$.

2.3.2. *Saturation effect.* The non linear effect of magnetic field saturation, modelled here, is directly linked to the hysteresis effect. When the magnetization field in the material reaches a certain value, given for example by Bozorth [5], this field cannot increase beyond this saturation value and becomes stationary. In the calculations, if the saturation value, denoted by M_{sat}, is exceeded, so that $\|\vec{M}\| > M_{sat}$, each component of the magnetization vector field is then proportionally modified, so that $\|\vec{M}\| = M_{sat}$.

2.4. Transformation of body forces into surface stresses

Let us consider the coordinate system shown in Figure 1a and used for the demonstration.

As said in the introduction, the three electromagnetic forces appearing as source terms in the elastic wave equation:

$$-\mu_L \Delta\vec{u} - (\lambda_L + \mu_L)\vec{\nabla}(\overline{\nabla}\vec{u}) + \rho\frac{\partial^2\vec{u}}{\partial t^2} = \overrightarrow{F_L} + \overrightarrow{F_M} + \overrightarrow{F_{MS}} \;, \tag{17}$$

have to be transformed into surface stresses. In equation (17), μ_L and λ_L are the Lamé's coefficients [Pa], $\vec{u}$ is the particle displacement [m] and ρ the material mass density [kg.m^{-3}].

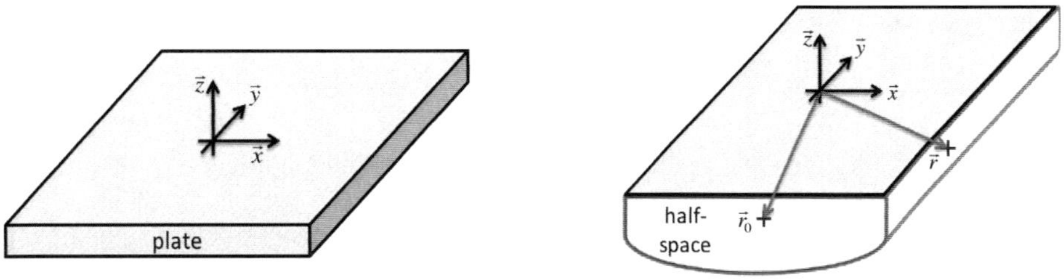

a) b)

Figure 1. Coordinate system for the transformation of body forces into surface stresses for a) the plate system and b) the half-space system where the point-like sources and the observation points are noticed $\overrightarrow{r_0}$ and $\vec{r}$, respectively.

As explained by Thompson [3], this transformation starts from a general formula for the particle displacement written as a spatial convolution of the Green's tensor $\bar{\bar{G}}$ with the sum of the body forces denoted by $\vec{f}$:

$$u_i(\vec{r}) = \sum_j \int_{-\infty}^{\infty}\int_{-\infty}^{\infty}\int_{-\infty}^{0} G_{ij}(\vec{r},\overrightarrow{r_0}) f_j(\overrightarrow{r_0}) dx_0 dy_0 dz_0 \;. \tag{18}$$

From this formula, results are sought to be rewritten as a surface convolution of the Green's functions with a distribution of equivalent surface stresses, as:

$$u_i(\vec{r}) = \sum_j \int_{-\infty}^{\infty}\int_{-\infty}^{\infty} \sigma_j^{\text{éq}}(\overrightarrow{r_0}|_{z_0=0}) G_{ij}(\vec{r},\overrightarrow{r_0}|_{z_0=0}) dx_0 dy_0 \;, \tag{19}$$

where $\vec{r} = (x, y, z)$ and $\overrightarrow{r_0} = (x_0, y_0, z_0)$ are the observation point and the source point, respectively. For convenience, the notation $\overrightarrow{r_0}|_{z_0=0}$ reduces to $\overrightarrow{R_0}$ in what follows.

The first transformation step consists in decomposing the Green's function into a Taylor series at the surface of a half-space, Figure 1b. This can be done because the spatial variations of the Green's functions are weak compared to those of the body forces: the former are related to the ultrasonic wavelength, typically of the order of few millimetres, while the latter are related to the skin depth, typically not exceeding few hundredths of a millimetre in ferromagnetic materials at frequencies used in EMAT applications. The Taylor series decomposition is stopped at the second order. Equation (18) becomes:

$$u_i(\vec{r}) = \sum_j \int_{-\infty}^{\infty}\int_{-\infty}^{\infty} \left[\begin{array}{c} G_{ij}(\vec{r},\overrightarrow{R_0}) T_j(\overrightarrow{R_0}) \\ + \dfrac{\partial G_{ij}(\vec{r},\overrightarrow{R_0})}{\partial z_0} M_j(\overrightarrow{R_0}) + \dfrac{1}{2}\dfrac{\partial^2 G_{ij}(\vec{r},\overrightarrow{R_0})}{\partial z_0^2} M_j(\overrightarrow{R_0}) \end{array} \right] dx_0 dy_0 \,, \tag{20}$$

where T_j, M_j and N_j represent the moments of the force f_j and are given by the following integrals:

$$T_j(\overrightarrow{R_0}) = \int_{-\infty}^{0} f_j(\overrightarrow{r_0})dz_0 \ , \ M_j(\overrightarrow{R_0}) = \int_{-\infty}^{0} z_0 f_j(\overrightarrow{r_0})dz_0 \ , \ N_j(\overrightarrow{R_0}) = \int_{-\infty}^{0} z_0^2 f_j(\overrightarrow{r_0})dz_0 \ . \tag{21}$$

The second step aims at eliminating the derivatives of the Green's functions over the thickness which appear in equation (20). Using the deformation tensor definition and the isotropic elasticity of the material, the local equilibrium law yields:

$$\sum_j K_{ij}(\overrightarrow{r_0})u_j(\overrightarrow{r_0}) = f_i(\overrightarrow{r_0}) \ , \tag{22}$$

with:

$$K_{ij}(\overrightarrow{r_0}) = \begin{pmatrix} -\rho\omega^2 - C_{11}\frac{\partial^2}{\partial x_0^2} - C_{44}\left(\frac{\partial^2}{\partial y_0^2}+\frac{\partial^2}{\partial z_0^2}\right) & -(C_{12}+C_{44})\frac{\partial^2}{\partial x_0\partial y_0} & -(C_{12}+C_{44})\frac{\partial^2}{\partial x_0\partial z_0} \\ -(C_{12}+C_{44})\frac{\partial^2}{\partial x_0\partial y_0} & -\rho\omega^2 - C_{11}\frac{\partial^2}{\partial y_0^2} - C_{44}\left(\frac{\partial^2}{\partial x_0^2}+\frac{\partial^2}{\partial z_0^2}\right) & -(C_{12}+C_{44})\frac{\partial^2}{\partial y_0\partial z_0} \\ -(C_{12}+C_{44})\frac{\partial^2}{\partial x_0\partial z_0} & -(C_{12}+C_{44})\frac{\partial^2}{\partial y_0\partial z_0} & -\rho\omega^2 - C_{11}\frac{\partial^2}{\partial z_0^2} - C_{44}\left(\frac{\partial^2}{\partial x_0^2}+\frac{\partial^2}{\partial y_0^2}\right) \end{pmatrix} \tag{23}$$

By multiplying equation (22) by G_{ki}, summing over i, integrating over the space and comparing the result with equation (18), we find:

$$u_k(\vec{r}) = \sum_{i,j} \int_{-\infty}^{\infty}\int_{-\infty}^{\infty}\int_{-\infty}^{0} G_{ki}(\vec{r},\overrightarrow{r_0})K_{ij}(\overrightarrow{r_0})u_j(\overrightarrow{r_0})\, dx_0 dy_0 dz_0 \ . \tag{24}$$

Considering the free surface condition and the energy conservation which imply that both $\bar{\bar{G}}$ and $\vec{u}$ tend to zero at infinity, the double integration by parts of the right-hand side of equation (24) gives:

$$u_k(\vec{r}) = \sum_{i,j}\int_{-\infty}^{\infty}\int_{-\infty}^{\infty} \begin{bmatrix} \int_{-\infty}^{0} u_j(\overrightarrow{r_0})K_{ij}(\overrightarrow{r_0})G_{ki}(\vec{r},\overrightarrow{r_0})dz_0 \\ +u_x(\overrightarrow{R_0})C_{44}\left[\frac{\partial G_{kz}(\vec{r},\overrightarrow{R_0})}{\partial x_0} + \frac{\partial G_{kx}(\vec{r},\overrightarrow{R_0})}{\partial z_0}\right] \\ +u_y(\overrightarrow{R_0})C_{44}\left[\frac{\partial G_{ky}(\vec{r},\overrightarrow{R_0})}{\partial z_0} + \frac{\partial G_{kz}(\vec{r},\overrightarrow{R_0})}{\partial y_0}\right] \\ +u_z(\overrightarrow{R_0})\left[C_{11}\frac{\partial G_{kz}(\vec{r},\overrightarrow{R_0})}{\partial z_0} + C_{12}\left(\frac{\partial G_{kx}(\vec{r},\overrightarrow{R_0})}{\partial x_0} + \frac{\partial G_{ky}(\vec{r},\overrightarrow{R_0})}{\partial y_0}\right)\right] \end{bmatrix} dx_0 dy_0 \ . \tag{25}$$

From this equation, the conditions of existence of the Green's functions are written as:

$$\sum_i K_{ij}(\overrightarrow{r_0})G_{ki}(\vec{r},\overrightarrow{r_0}) = \delta^3(\vec{r}-\overrightarrow{r_0})\delta_{kj} \ , \tag{26}$$

$$\frac{\partial G_{kz}(\vec{r},\overrightarrow{R_0})}{\partial x_0} + \frac{\partial G_{kx}(\vec{r},\overrightarrow{R_0})}{\partial z_0} = 0 \ , \tag{27}$$

$$\frac{\partial G_{ky}(\vec{r},\overrightarrow{R_0})}{\partial z_0} + \frac{\partial G_{kz}(\vec{r},\overrightarrow{R_0})}{\partial y_0} = 0 \ , \tag{28}$$

$$C_{11}\frac{\partial G_{kz}(\vec{r},\overrightarrow{R_0})}{\partial z_0} + C_{12}\left[\frac{\partial G_{kx}(\vec{r},\overrightarrow{R_0})}{\partial x_0} + \frac{\partial G_{ky}(\vec{r},\overrightarrow{R_0})}{\partial y_0}\right] = 0 \ , \tag{29}$$

where δ_{kj} denotes the Kronecker symbol. Using the development of equation (26) with the expression of the tensor $\bar{\bar{K}}$, equation (23), and the derivatives of equations (27) to (29), the first order and the second order of the derivatives in z_0 of the Green's functions are given by:

$$\frac{\partial G_{kz}(\vec{r},\overrightarrow{R_0})}{\partial z_0} = -\frac{C_{12}}{C_{11}}\left[\frac{\partial G_{kx}(\vec{r},\overrightarrow{R_0})}{\partial x_0} + \frac{\partial G_{ky}(\vec{r},\overrightarrow{R_0})}{\partial y_0}\right], \tag{30}$$

$$\frac{\partial G_{kx}(\vec{r},\overrightarrow{R_0})}{\partial z_0} = -\frac{\partial G_{kz}(\vec{r},\overrightarrow{R_0})}{\partial x_0}, \tag{31}$$

$$\frac{\partial G_{ky}(\vec{r},\overrightarrow{R_0})}{\partial z_0} = -\frac{\partial G_{kz}(\vec{r},\overrightarrow{R_0})}{\partial y_0}, \tag{32}$$

$$\frac{\partial^2 G_{kz}(\vec{r},\overrightarrow{R_0})}{\partial z_0^2} = \frac{C_{12}}{C_{11}}\left[\frac{\rho\omega^2}{C_{44}} + \frac{\partial^2}{\partial x_0^2} + \frac{\partial^2}{\partial y_0^2}\right] G_{kz}(\vec{r},\overrightarrow{R_0}), \tag{33}$$

$$\frac{\partial^2 G_{kx}(\vec{r},\overrightarrow{R_0})}{\partial z_0^2} = -\left[\frac{\rho\omega^2}{C_{44}} + \left(2+\frac{C_{12}}{C_{11}}\right)\frac{\partial^2}{\partial x_0^2} + \frac{\partial^2}{\partial y_0^2}\right] G_{ky}(\vec{r},\overrightarrow{R_0}) - \left(1+\frac{C_{12}}{C_{11}}\right)\frac{\partial^2}{\partial x_0 \partial y_0} G_{ky}(\vec{r},\overrightarrow{R_0}), \tag{34}$$

$$\frac{\partial^2 G_{ky}(\vec{r},\overrightarrow{R_0})}{\partial z_0^2} = -\left[\frac{\rho\omega^2}{C_{44}} + \frac{\partial^2}{\partial x_0^2} + \left(2+\frac{C_{12}}{C_{11}}\right)\frac{\partial^2}{\partial y_0^2}\right] G_{ky}(\vec{r},\overrightarrow{R_0}) - \left(1+\frac{C_{12}}{C_{11}}\right)\frac{\partial^2}{\partial x_0 \partial y_0} G_{kx}(\vec{r},\overrightarrow{R_0}), \tag{35}$$

Finally, substituting equations (30) to (35) into equation (21) yields:

$$u_i(\vec{r}) = \int_{-\infty}^{\infty}\int_{-\infty}^{\infty} \left[\begin{array}{c} \left[T_x(\overrightarrow{R_0}) - \frac{1}{2}\frac{\rho\omega^2}{C_{44}}N_x(\overrightarrow{R_0})\right] G_{ix}(\vec{r},\overrightarrow{R_0}) - \frac{C_{12}}{C_{11}}M_z(\overrightarrow{R_0})\frac{\partial}{\partial x_0}G_{ix}(\vec{r},\overrightarrow{R_0}) \\ -\frac{1}{2}\left[N_x(\overrightarrow{R_0})\left[\left(2+\frac{C_{12}}{C_{11}}\right)\frac{\partial^2}{\partial x_0^2} + \frac{\partial^2}{\partial y_0^2}\right] + N_y(\overrightarrow{R_0})\left(1+\frac{C_{12}}{C_{11}}\right)\frac{\partial^2}{\partial x_0\partial y_0}\right] G_{ix}(\vec{r},\overrightarrow{R_0}) \\ + \left[T_y(\overrightarrow{R_0}) - \frac{1}{2}\frac{\rho\omega^2}{C_{44}}N_y(\overrightarrow{R_0})\right] G_{iy}(\vec{r},\overrightarrow{R_0}) - \frac{C_{12}}{C_{11}}M_z(\overrightarrow{R_0})\frac{\partial}{\partial y_0}G_{iy}(\vec{r},\overrightarrow{R_0}) \\ -\frac{1}{2}\left[N_x(\overrightarrow{R_0})\left(1+\frac{C_{12}}{C_{11}}\right)\frac{\partial^2}{\partial x_0\partial y_0} + N_y(\overrightarrow{R_0})\left[\frac{\partial^2}{\partial x_0^2} + \left(2+\frac{C_{12}}{C_{11}}\right)\frac{\partial^2}{\partial y_0^2}\right]\right] G_{iy}(\vec{r},\overrightarrow{R_0}) \\ + \left[T_z(\overrightarrow{R_0}) + \frac{1}{2}\frac{\rho\omega^2}{C_{44}}\frac{C_{12}}{C_{11}}N_z(\overrightarrow{R_0})\right] G_{iz}(\vec{r},\overrightarrow{R_0}) \\ -\left[M_x(\overrightarrow{R_0})\frac{\partial}{\partial x_0} + M_y(\overrightarrow{R_0})\frac{\partial}{\partial y_0}\right] G_{iz}(\vec{r},\overrightarrow{R_0}) \\ + \frac{1}{2}\frac{C_{12}}{C_{11}}N_z(\overrightarrow{R_0})\left[\frac{\partial^2}{\partial x_0^2} + \frac{\partial^2}{\partial y_0^2}\right] G_{iz}(\vec{r},\overrightarrow{R_0}) \end{array}\right] dx_0 dy_0 \,. \tag{36}$$

The equivalent surface stresses appearing in equation (19) are readily identified as:

$$\sigma_x^{eq} = T_x - \frac{1}{2}\frac{\rho\omega^2}{C_{44}}N_x - \frac{C_{12}}{C_{11}}M_z\frac{\partial}{\partial x_0} - \frac{1}{2}\left[N_x\left[\left(2+\frac{C_{12}}{C_{11}}\right)\frac{\partial^2}{\partial x_0^2} + \frac{\partial^2}{\partial y_0^2}\right] + N_y\left(1+\frac{C_{12}}{C_{11}}\right)\frac{\partial^2}{\partial x_0\partial y_0}\right], \tag{37}$$

$$\sigma_y^{eq} = T_y - \frac{1}{2}\frac{\rho\omega^2}{C_{44}}N_y - \frac{C_{12}}{C_{11}}M_z\frac{\partial}{\partial y_0} - \frac{1}{2}\left[N_x\left(1+\frac{C_{12}}{C_{11}}\right)\frac{\partial^2}{\partial x_0\partial y_0} + N_y\left[\frac{\partial^2}{\partial x_0^2} + \left(2+\frac{C_{12}}{C_{11}}\right)\frac{\partial^2}{\partial y_0^2}\right]\right], \tag{38}$$

$$\sigma_z^{eq} = T_z + \frac{1}{2}\frac{\rho\omega^2}{C_{44}}\frac{C_{12}}{C_{11}}N_z - M_x\frac{\partial}{\partial x_0} - M_y\frac{\partial}{\partial y_0} + \frac{1}{2}\frac{C_{12}}{C_{11}}N_z\left[\frac{\partial^2}{\partial x_0^2} + \frac{\partial^2}{\partial y_0^2}\right]. \tag{39}$$

Equations (37-39) constitute a general result though they were derived to deal with typical forces generated by EMAT. Each term depends on the various components of the force considered as it depends of the force moments.

It must be noticed that the basic assumptions and approximations made in deriving them depend only on elastodynamic quantities. Therefore, these formulae can be used to transform a distribution of body forces of any physical origin into an equivalent distribution of surface stresses; the main condition of applicability is that body forces must be concentrated in the vicinity of the free surface of an elastic medium. The notion of vicinity is understood as follows: whatever the direction of the vector forces, the extent of the distribution of forces in the direction normal to the free surface must be small compared to the elastic wavelength.

Note that a result based on similar assumptions and approximations was proposed in [4] in the more restrictive case of a 2D distribution of horizontal forces assumed to be normal to the 2D plane of elastic wave propagation, therefore limiting the applicability to sources of shear horizontal waves.

3. Example of application in 2D: radiation of Lamb waves by a meander-coil EMAT

The coordinate system is again that of Figure 1a but in 2D, *i.e.*, the $\vec{y}$ axis is the infinite dimension along which derivatives vanish.

3.1. Green's function for Lamb waves generated in a plate by a surface stress

The Green's functions appearing in equation (36) are not explicitly given. Here, 2D Green's functions for a line source at the surface of an isotropic plate is given in the form of a modal decomposition, well suited for further uses of NDT simulation [10], the knowledge of modal amplitudes allowing easy interpretation of simulated results. They are derived using the reciprocity relation given by Auld [11].

Let $A_m^{\sigma_{xz}}$ and $A_m^{\sigma_{zz}}$ denote the amplitudes associated with the Lamb mode m due to a tangential stress σ_{xz} and to a normal stress σ_{zz}, respectively. Their calculations are based on the integration over the plate thickness of the complex reciprocity relation [2]. Equivalent results were obtained for the shear horizontal case in [4] and a similar derivation is made here to get the modal amplitudes for the Lamb case. Amplitudes are:

$$A_m^{\sigma_{xz}}(x, x_0) = \frac{e^{-j\beta_m x}}{4P_{mm}} \tilde{v}_{x_m}(z = 0)e^{j\beta_m x_0} \ , \tag{40}$$

$$A_m^{\sigma_{zz}}(x, x_0) = \frac{e^{-j\beta_m x}}{4P_{mm}} \tilde{v}_{z_m}(z = 0)e^{j\beta_m x_0} \ , \tag{41}$$

where x represents the observation point, x_0 the position of the line source, β_m the wavenumber of the m-th mode; $\tilde{v}_{x_m}$ and $\tilde{v}_{z_m}$ are the components of the particle velocity of the m-th mode taken at the plate surface. The factor P_{mm} is given by:

$$P_{mm} = -\frac{1}{4}\int_{-h}^{0} \left(\vec{v}_m^{*}\bar{\bar{\sigma}}_m + \vec{v}_m\bar{\bar{\sigma}}_m^{*}\right).\vec{x}\, dz \ , \tag{40}$$

where $\bar{\bar{\sigma}}_m$ stands for the amplitude of the stress tensor of the m-th mode and h the plate thickness. The notation * denotes the complex-conjugate value of the designated quantity. The wavenumbers β_m are analytically known for shear horizontal (SH) waves but must be numerically evaluated in the case of Lamb waves. For this, our implementation of the semi analytical finite element method is used.

3.2. Modal amplitude of Lamb guided modes generated by an EMAT

By assuming the previous modal decomposition, the displacement $\vec{u_G}$ generated by a line source of normal or tangential stresses over a plate is written, respectively:

$$\vec{u}_G^{\sigma_{xz}}(x, [x_0], z) = \sum_m A_m^{\sigma_{xz}}(x, x_0)\,\vec{u}_m(z) = \sum_m \frac{e^{-j\beta_m x}}{4P_{mm}} \tilde{v}_{x_m}(z = 0)e^{j\beta_m x_0}\vec{\tilde{u}}_m(z) \ , \tag{41}$$

$$\vec{u}_G^{\sigma_{zz}}(x, [x_0], z) = \sum_m A_m^{\sigma_{zz}}(x, x_0)\,\vec{u}_m(z) = \sum_m \frac{e^{-j\beta_m x}}{4P_{mm}} \tilde{v}_{z_m}(z = 0)e^{j\beta_m x_0}\vec{\tilde{u}}_m(z) \ , \tag{42}$$

The expression (18) of the particle displacement uses Green's functions for a half space which verifies the free surface conditions. The Green's functions developed in the previous paragraph verify similar conditions at both plate surfaces. However, at the surface opposite to that where the source applies, the generated body forces are negligible as soon as the condition "plate thickness much larger than the skin depth" is respected. This is generally true at typical frequencies and for the materials considered in NDT applications. Therefore, the half-space Green's functions may be substituted with the elastic plate Green's functions.

For a transducer of length L, a derivation similar to that made in the simpler case of SH waves in [4] gives for a force f_x or f_z, respectively:

$$A_m^{fx}(x) = \frac{e^{-j\beta_m x}}{4P_{mm}} \tilde{v}_{x_m}(z=0) \int_{-L/2}^{L/2} \left[T_x - j\beta_m M_x - \frac{1}{2}\frac{\rho\omega^2}{C_{44}} N_x + \frac{1}{2}\left(2 + \frac{C_{12}}{C_{11}}\right)\beta_m^2 N_x\right] e^{j\beta_m x_0} dx_0 \,, \quad (43)$$

$$A_m^{fz}(x) = \frac{e^{-j\beta_m x}}{4P_{mm}} \tilde{v}_{z_m}(z=0) \int_{-L/2}^{L/2} \left[T_z - j\frac{C_{12}}{C_{11}}\beta_m M_z + \frac{1}{2}\frac{\rho\omega^2}{C_{44}}\frac{C_{12}}{C_{11}} N_z - \frac{1}{2}\frac{C_{12}}{C_{11}}\beta_m^2 N_z\right] e^{j\beta_m x_0} dx_0 \,. \quad (44)$$

These expressions derive from equations (37-39) when considering the simplifications resulting from the 2D configuration considered in this part.

3.3. Meander-coil EMAT configuration

The theory described in previous paragraphs is now applied to model a meander-coil EMAT operating on a 3-mm-thick iron plate. The configuration and the parameters used in the computations are first presented. All the data are given at room temperature and the temperature dependence is not considered.

Iron is modelled as elastic and isotropic with mass density of 7700 kg.m^{-3}, longitudinal velocity of 5690 m.s^{-1} and shear velocity of 3145 m.s^{-1}. The static magnetic permeability and the dynamic are set to 200 and 1000 respectively, and the magnetic susceptibility is 200. The electric conductivity is 9.93 MS.m^{-1}. The magnetization at saturation is 1720 kA.m^{-1}. The magnetostriction curve is analytically approximated by Hirao and Ogi [8]. The EMAT used is schematized on Figure 2. The meander-coil is made of 10 wires of diameter a = 1 mm, and the inter-space between two wires is c = 0.5 mm, so that its total length L equals 15 mm. The lift-off is 0.1 mm and the induction current sent to the coil is 1 mA with a frequency varying in the [0.1; 5] MHz range by 10 kHz steps (491 calculation frequencies).

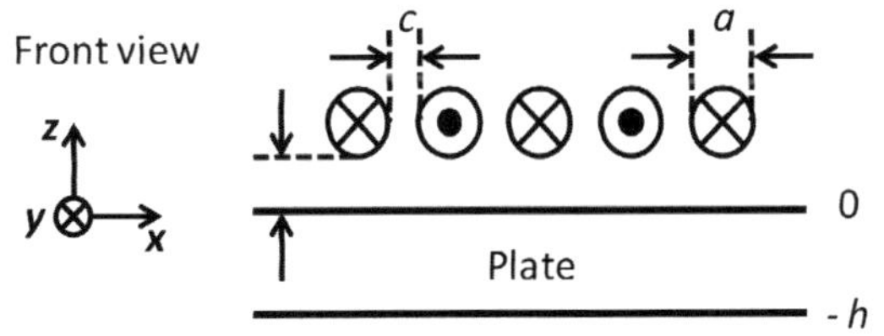

Figure 2. Representation of the meander-coil EMAT used in simulations.

The magnetic fields, and thus, the three electromagnetic forces, are computed at each point of a grid in the plate. The length of the transducer (in the $\vec{x}$ direction) is discretized by 150 regularly-spaced points (*i.e.* 10 points per wire and 5 points between two wires). The thickness of the material (in the $\vec{z}$ direction) is divided into two regions: the first region from the surface to a depth equals to five times the skin depth is discretized by 50 points; the remaining thickness is meshed by only 5 points, because the fields, thus, the forces, are confined close to the surface.

Before calculating the modal amplitudes, it is important to check if the assumptions made in deriving the model are satisfied. At first, the static field set at 0.5 T is much larger than the dynamic one which order of magnitude is 10^{-5} T for the induction current considered. Then, discretizations in space and time are chosen so that the assumptions made are fulfilled.

3.4. Results

The modal amplitudes of SH and Lamb waves calculated are presented for different static magnetization directions, $\vec{x}$, $\vec{y}$ and $\vec{z}$. The static field is considered uniform under the material and is set to 0.5 T in the three system directions, successively.

Results shown in figure 3 are presented to illustrate the capability of our model to predict elastic wave fields radiated by EMAT of various designs. Here, the three possible orientations of the static field relatively to the wire direction of a meander-coil EMAT are considered. Several observations can be made from these results.

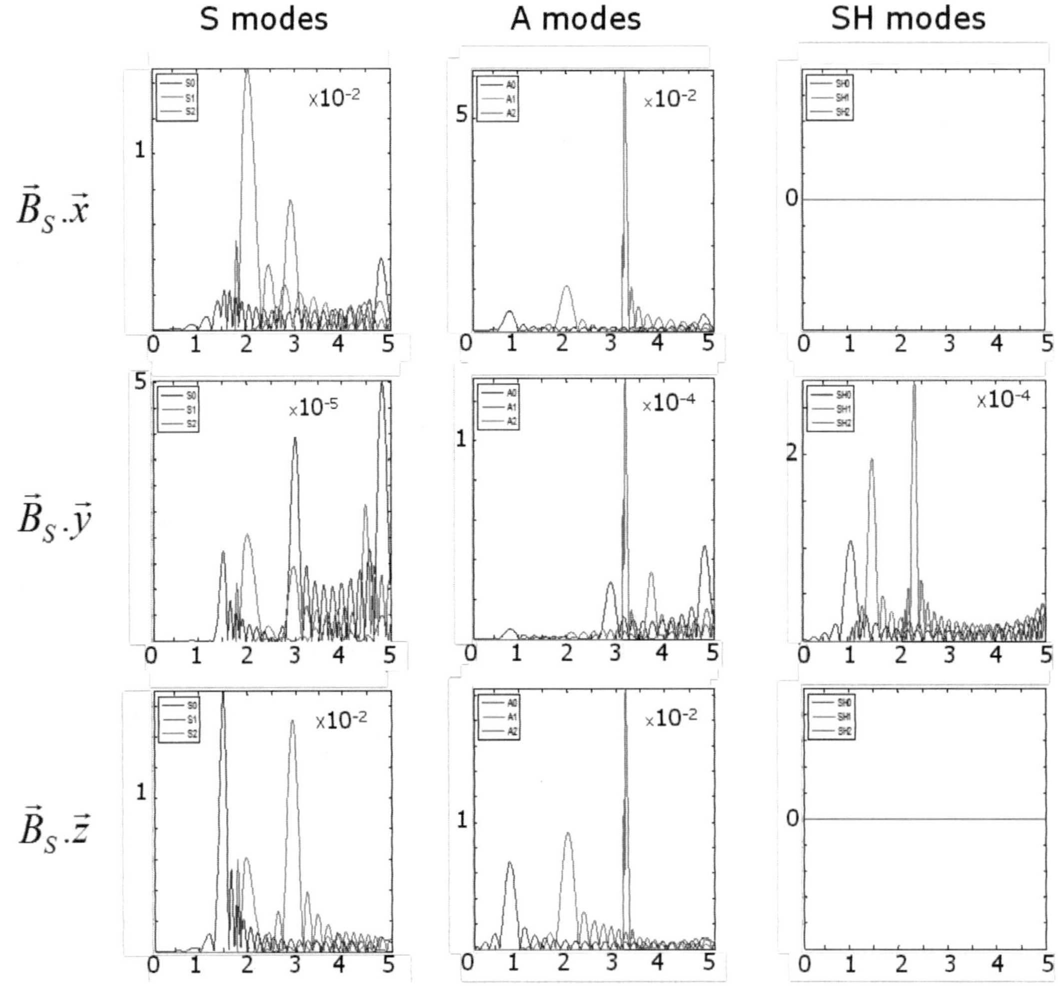

Figure 3. Modal amplitude over the frequency [MHz] of the first three symmetric (S), antisymmetric (A) and shear horizontal (SH) modes for three different directions of the permanent magnet static field, $\vec{x}$, $\vec{y}$ and $\vec{z}$ in a 3 mm thick-iron plate.

First, a modulation is visible in the amplitudes displayed, whatever the static field orientation. It is related to the coil design as the peak values arise when the wire spacing of the coil coincides with the wavelength of the mode. Then, shear horizontal (SH) modes are generated only when the static field direction coincides with the electric current direction: SH waves result from forces acting along the $\vec{y}$ axis which only exist when the static field and the electric current directions are aligned (in the 2D simulations considered here). Interestingly, in this case, the amplitude of SH waves can reach higher values than that of symmetric (S) and antisymmetric (A) Lamb modes. However, the relative amplitudes of the various modal contributions depend on the excitation frequency.

Numerous results were obtained but it is not the purpose of the present paper to discuss them. Many parameters can be varied in our modelling approach, even in the simple 2D configurations that are treated. Depending on the material characteristics, one of the three forces may dominate the others.

Note that 2D configurations correspond to actual cases of EMAT use in NDT applications. Experimental studies will be performed in order to validate the theoretical model derived here.

4. Conclusions

The magnetic fields and the electromagnetic forces generated by an EMAT in a ferromagnetic material have been described. The double frequency effect was highlighted. The three generated body forces were then transformed into equivalent surface stresses and the role of the Green's functions involved in this transformation has been described. Under this transformation, the modal amplitudes of Lamb waves generated by an EMAT in a ferromagnetic plate have been expressed as quasi-closed form solutions; these solutions complement our earlier closed form solutions restricted to SH waves. Examples of application have been given to illustrate the capabilities of our model.

It must be noticed that a similar transformation could be applied to the case of the radiation of bulk or surface waves by simply adopting the appropriate Green's functions to describe their specific propagation characteristics.

To apply the presented theory, some assumptions must be verified. However, they are not restrictive in regards to typical EMAT designs and to the characteristics of ferromagnetic materials of interest in NDT applications.

These analytical (for SH waves) and quasi-analytical (for Lamb waves) calculations allow fast parametric studies for optimizing EMAT design for a given material.

Various stages in the model derivation could be adapted to get an overall model which could apply in more general cases, such adaptations being compatible with the overall approach proposed herein. For examples, the magnetization field can be calculated from the Weiss molecular field theory and the magnetic permeability can be expressed as a function of the applied field amplitude and the induction frequency. The magnetostriction model can also be improved, especially in relating the magnetostrictive deformation calculated by Bozorth [5] and the magnetostrictive strain $\bar{\bar{\varepsilon}}$, by considering for example a micro-macro homogenization model [12].

References

[1] Thompson R B 1990 Physical principles of measurements with EMAT transducers *Physical Acoustics* vol 19 (New-York: Academic Press) chapter 4 pp 157–200

[2] Ditri J J and Rose J F 1992 *J. Appl. Phys.* **72** 2589–97

[3] Thompson R B 1980 *J. Nond. Eval.* **1** 79–85

[4] Rouge C, Lhémery A and Ségur D 2012 *J. Phys.: Conf. Ser.* **353** 012014

[5] Bozorth R M 2003 *Ferromagnetism* (Hoboken: Wiley-Interscience)

[6] Bowler N 2006 Frequency-depenence of relative permeability in steel *Review of Progress in QNDE* **25** , ed D O Thompson and D E Chimenti (Melville: AIP Conf. Proc. 820) 1269

[7] Ribichini R, Cegla F, Nagy P B and Cawley P 2010 *IEEE Trans. Ultra. Ferro. Freq. Cont.* **57** 2808–17

[8] Hirao M and Ogi H 2003 *EMATS for Science and Industry: Noncontacting Ultrasonic Measurements* (Boston: Kluwer Academic Publishers)

[9] Laguerre L, Aime J-C and Brissaud M 2002 *Ultrasonics* **39** 503–14

[10] Jezzine K and Lhémery A 2007 Simulation of guided wave inspection based on the reciprocity principle and the semi-analytical finite element method *Review of Progress in QNDE* **26**, ed D O Thompson and D E Chimenti (Melville: AIP Conf. Proc. 894) 39

[11] Auld B A 1973 *Acoustics fields and waves in solids* vol 2 (Malabar, Florida: Krieger publishing company)

[12] Daniel L, Hubert O, Buiron N and Billardon R 2008 *J. Mech. Phys. Solids* **56** 1018–42

Frequency spectra of magnetostrictive and Lorentz forces generated in ferromagnetic materials by a CW excited EMAT

C Rouge[1,2]**, A Lhémery**[1] **and C Aristégui**[2]

[1] CEA, LIST, F-91191 Gif-sur-Yvette cedex, France
[2] Université de Bordeaux, UMR CNRS 5295, I2M/Acoustique Physique, 33405 Talence cedex, France

E-mail: alain.lhemery@cea.fr

Abstract. Magnetostriction arises in ferromagnetic materials subjected to magnetization, *e.g.*, when an EMAT (Electro-Magnetic Acoustic Transducer) is used to generate ultrasonic waves. In such a case, the magnetostriction force must be taken into account as a transduction process that adds up to the Lorentz force. When the static magnetic field is high compared to the dynamic field, both forces are driven by the excitation frequency. For lower static relative values of the magnetic fields, the Lorentz force comprises both the excitation frequency and its first harmonic. In this work, a model is derived to predict the frequency content of the magnetostrictive force that comprises several harmonics. The discrete frequency spectrum strongly depends on both the static field and the relative amplitude of the dynamic field. The only material input data needed to predict it is the curve of macroscopic magnetostrictive strain that can be measured in the direction of an imposed magnetic field. Then, the various frequency-dependent distributions of Lorentz and magnetostriction body forces can be transformed into equivalent surface stresses. Examples of computation are given for different static and dynamic magnetic fields to study their influence on the frequency content of waves generated in ferromagnetic materials.

1. Introduction

Electro-magnetic acoustic transducers (EMAT) are commonly used in the context of ultrasonic nondestructive testing. They are often used to generate guided waves and more specifically shear horizontal waves in plates or torsional waves in cylinders. In some cases, EMAT efficiency is increased when they are used with magnetostrictive patches or in ferromagnetic materials [1]. The two main processes at the origin of wave generation by an EMAT are the Lorentz and the magnetostriction forces [2, 3]. The first exists in all conductive media while the second is restrictive to ferromagnetic materials. The two forces can be the source of ultrasounds at frequencies different from the excitation frequency. For the Lorentz force, the excitation of waves at a frequency twice the excitation frequency is easily understood, as recalled hereafter. This is more complicated dealing with the magnetostriction force; its frequency spectrum cannot be explicitly given.

The first aim of present work is to model the frequency spectrum of the magnetostriction force. In addition, to compute the amplitude of waves generated by an EMAT in a ferromagnetic material, the magnetostriction and the Lorentz forces are transformed into surface stresses that can be readily used as source terms in existing semi-analytical models of wave radiation of bulk [4] or guided waves [5]. The transformation explained in two previous papers [6, 7] is valid since the two body forces are constrained in the vicinity of the plate surface. The order of magnitude of the equivalent surface stress of the magnetostriction force is compared to results from the literature. A very good estimation of this quantity is obtained as compared to Ribichini *et al.* [8] finite element results. Then, the contributions of the fundamental and the first four harmonics of the magnetostriction force to SH0 guided mode in a ferromagnetic plate are given as an example of application. The fact that harmonic amplitudes cannot be always neglected is clearly illustrated.

2. Frequency spectra of the electromagnetic forces

2.1. Magnetic field created by an EMAT

The Lorentz force $\mathbf{F}_\mathrm{L}$ results from the interaction of eddy currents $\mathbf{J}$ created in the material by the EMAT coil and the magnetic induction field $\mathbf{B}$ generated by both the permanent magnet and the coil. It is expressed as follows

$$\mathbf{F}_\mathrm{L} = \mathbf{J}\times\mathbf{B}\,. \tag{1}$$

If the magnetic induction field is decomposed into a static field $\mathbf{B}_\mathrm{S}$ and a dynamic one $\mathbf{B}_\mathrm{D}\,e^{j\omega_0 t}$, the Lorentz force can be re-written as

$$\mathbf{F}_\mathrm{L} = \mathbf{J}_\mathrm{D}\times\mathbf{B}_\mathrm{S}e^{j\omega_0 t} + \mathbf{J}_\mathrm{D}\times\mathbf{B}_\mathrm{D}e^{2j\omega_0 t}\,, \tag{2}$$

where eddy currents are pure dynamic quantities with $\mathbf{J} = \mathbf{J}_\mathrm{D}\,e^{j\omega_0 t}$ and where a harmonic excitation of angular frequency ω_0 is considered. This expression highlights the two interactions at its origin. The first is that of eddy currents generated in the material by the coil with the static magnetic field due to the permanent magnet; it is given by the first term of the right hand side of equation (2) and involves a $e^{j\omega_0 t}$ factor. Eddy currents also interact with the dynamic magnetic field; this is expressed by the second term proportional to $e^{2j\omega_0 t}$. If the dynamic magnetic field is not negligible compared to the static one, the double frequency effect appears.

2.2. Calculation of the magnetostrictive strain coefficients

As formulated by Thompson [1] and Ogi [2], the calculation of the magnetostriction force reduces to the evaluation of the magnetostrictive strain coefficients. Two 2D cases have to be considered in order to form the complete magnetostrictive strain tensor in 3D.

2.2.1. Dynamic magnetic field collinear with the static magnetic field. The first case to consider is when the static magnetic field $\mathbf{H}_\mathrm{S}$ and the dynamic one $\mathbf{H}_\mathrm{D}$ are collinear, as shown in Figure 1 (left). The static field induces a static magnetostrictive strain.

This fixed value is shown in Figure 1 (right) on the magnetostriction curve $\varepsilon(H)$ [2] that represents the magnetostrictive strain due to a magnetic field. This curve shown is typical of steels. A superimposed dynamic field produces time-dependent oscillations of the magnetostrictive strain around this fixed value. Thus, the total excitation field $\mathbf{H}_\mathrm{T}$ (static plus dynamic fields) induces a time-dependent magnetostrictive strain ε_T.

To express the magnetostrictive strain vector in a plane containing the two magnetic fields, two assumptions are made. The first one is that the magnetostriction curve, Figure 1 (right), is the same at any point of the material studied for any applied magnetic field direction. The second is that the magnetostriction effect keeps the material volume constant. Under these two assumptions, the magnetostrictive strain vector is written as

$$\boldsymbol{\varepsilon}_{\mathbf{MS}} = \begin{pmatrix} \varepsilon_T \\ -\dfrac{1}{2}\varepsilon_T \\ -\dfrac{1}{2}\varepsilon_T \\ 0 \\ 0 \\ 0 \end{pmatrix}. \qquad (3)$$

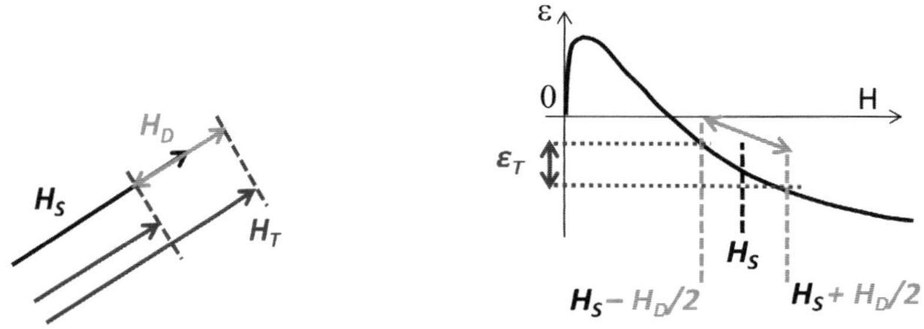

Figure 1: Case 1: the static and the dynamic magnetic fields are collinear (left); they induce different magnetostrictive strains represented on the magnetostriction curve (steels) (right).

These two assumptions are not restrictive in the context of NDT by EMAT in regards of typical materials tested and magnetic field amplitudes used. As far as the first is concerned, it is a consequence of magnetic isotropy that is observed in disordered materials. The second assumption can be made unless very high magnetization fields are generated; for example, it applies very well for a field of 0.5 T (see Bozorth [9]) which is already quite large for an EMAT used in NDT.

2.2.2. Dynamic magnetic field perpendicular to the static magnetic field. The case where the static and the dynamic excitation fields are not collinear is now considered. For example, the static field is aligned along the $\mathbf{x}$-axis and the dynamic one is perpendicular to it, as shown on Figure 2. The total field is thus applied in the resulting direction, denoted by $\mathbf{x}'$, which is time-dependent, making an angle θ with the static field direction.

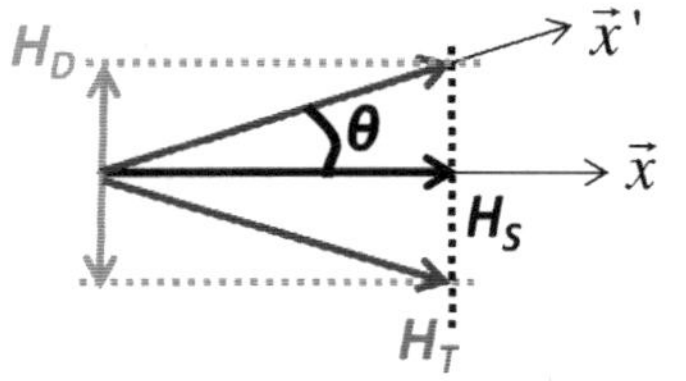

Figure 2: Case 2: the static and the dynamic magnetic fields are perpendicular.

Under the same assumptions as made in § 2.2.1, the magnetostrictive strain vector in the $\mathbf{x'}$ direction is expressed similarly as in equation (3), but with another total strain denoted by ε_T^{xy}. A rotation of angle θ must be further applied to express it in the absolute coordinates. One gets

$$\boldsymbol{\varepsilon}_{\mathbf{MS}} = \begin{pmatrix} \varepsilon_T^{xy}\cos^2(\theta) - \frac{1}{2}\varepsilon_T^{xy}\sin^2(\theta) \\ -\frac{1}{2}\varepsilon_T^{xy}\cos^2(\theta) + \varepsilon_T^{xy}\sin^2(\theta) \\ -\frac{1}{2}\varepsilon_T^{xy} \\ 0 \\ 0 \\ \frac{3}{2}\varepsilon_T^{xy}\sin(2\theta) \end{pmatrix}. \tag{4}$$

2.2.3. Magnetostrictive strain tensor. Finally, if the static field direction is along the **x** axis, the magnetostriction strain tensor is readily written as

$$\boldsymbol{\varepsilon}_{\mathbf{MS}} = \begin{pmatrix} \varepsilon_T & \varepsilon_T^{xy}\cos^2(\theta_{xy}) - \frac{1}{2}\varepsilon_T^{xy}\sin^2(\theta_{xy}) & \varepsilon_T^{xz}\cos^2(\theta_{xz}) - \frac{1}{2}\varepsilon_T^{xz}\sin^2(\theta_{xz}) \\ -\frac{1}{2}\varepsilon_T & -\frac{1}{2}\varepsilon_T^{xy}\cos^2(\theta_{xy}) + \varepsilon_T^{xy}\sin^2(\theta_{xy}) & -\frac{1}{2}\varepsilon_T^{xz} \\ -\frac{1}{2}\varepsilon_T & -\frac{1}{2}\varepsilon_T^{xy} & -\frac{1}{2}\varepsilon_T^{xz}\cos^2(\theta_{xz}) + \varepsilon_T^{xz}\sin^2(\theta_{xz}) \\ 0 & 0 & 0 \\ 0 & 0 & \frac{3}{2}\varepsilon_T^{xz}\sin(2\theta_{xz}) \\ 0 & \frac{3}{2}\varepsilon_T^{xy}\sin(2\theta_{xy}) & 0 \end{pmatrix}, \tag{5}$$

where the first column represents the magnetostrictive strain along the **x** axis with a total deformation in this direction denoted by ε_T; the second represents the magnetostrictive strain in the $(\mathbf{x},\mathbf{y})$ plane with a total deformation ε_T^{xy} in the direction of the total magnetic field and an angle θ_{xy} between the **x**-axis and the direction of the total magnetic field. Accordingly, the third represents the magnetostrictive strain in the $(\mathbf{x},\mathbf{z})$ plane with a total deformation ε_T^{xz} and an angle θ_{xz}. The various quantities appearing in equation (5) are time-dependent.

2.3. Magnetostriction force frequency spectrum

From the previous strain expression, the magnetostriction force can be calculated. Five steps are necessary. The first consists in calculating the piezomagnetic strain tensor $\mathbf{d}$, given by

$$d_{Ij} = \frac{\partial \varepsilon_{MS_I}}{\partial H_j}. \tag{6}$$

Next, the piezomagnetic stress tensor $\mathbf{e}$ is calculated with the help of the elastic stiffness tensor $\mathbf{C}$:

$$\mathbf{e} = \mathbf{C}\mathbf{d}. \tag{7}$$

Then, the magnetostrictive stress tensor $\boldsymbol{\sigma}_{\mathbf{MS}}$ is expressed by

$$\boldsymbol{\sigma}_{\mathbf{MS}} = -\mathbf{e}\mathbf{H}. \tag{8}$$

The stress free boundary conditions at the part surface are accounted for in the calculation of the magnetostrictive stress tensor by writing

$$\boldsymbol{\sigma}_{\mathbf{MS}} \cdot \mathbf{n} = 0. \tag{9}$$

Finally, the magnetostriction force $\mathbf{F}_{\mathbf{MS}}$ is expressed by

$$\mathbf{F}_{\mathbf{MS}} = \nabla \cdot \boldsymbol{\sigma}_{\mathbf{MS}} . \tag{10}$$

The frequency content of the magnetostriction force does not explicitly appear in this formulation. Its evaluation is made by means of the following algorithm, schematized by Figure 3.

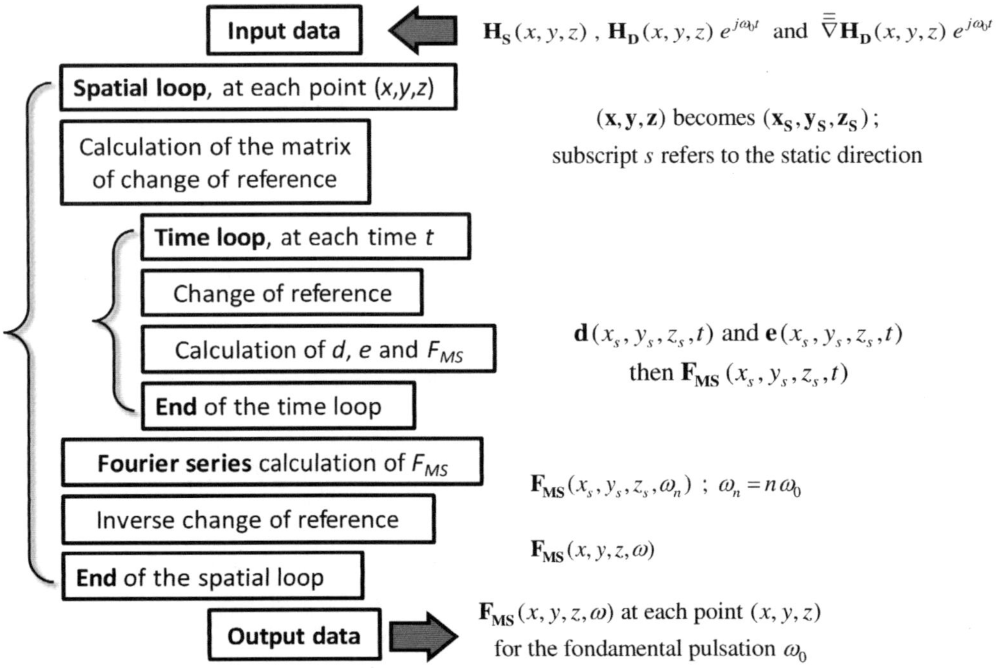

Figure 3: Calculation of the magnetostriction force frequency spectrum for a harmonic excitation ω_0.

For a given CW excitation of angular frequency ω_0, the magnetic fields $\mathbf{H_S}$ and $\mathbf{H_D}$ are computed. They constitute the input of the calculation. At each point (x, y, z) in the global reference system $(\mathbf{x},\mathbf{y},\mathbf{z})$, a local reference system $(\mathbf{x_S},\mathbf{y_S},\mathbf{z_S})$ is used: the first axis $\mathbf{x_S}$ coincides with the static field direction at the point considered. Previous expressions of $\boldsymbol{\varepsilon}_{\mathbf{MS}}$ can thus be readily used. Next, the magnetostriction force is calculated in the time domain. This is done in a time loop over one period of the excitation frequency. At the end of this time-loop, the frequency spectrum of the magnetostriction force is calculated thanks to a Fourier series, and is finally expressed in the global reference system. At the end of the algorithm, the frequency spectrum is computed at all the calculation points for a given excitation frequency.

2.4. Examples of spectra

Let us now consider examples of frequency spectra of magnetostriction force. The magnetostriction curve used is that previously shown in Figure 1 (right), [2].

Three static field intensities are considered corresponding respectively to a weak, a moderate and a high bias field induction of 0.5, 2 and 6.5 10^4 A.m^{-1}, respectively (see Figure 4). Three dynamic field intensities are also considered: 0.1, 0.5 and 1 10^4 A.m^{-1}. The frequency spectra of a point situated at the plate surface below the middle of a wire are represented in Figure 5. They are normalized with respect to their highest component. Results show a general behaviour already described in the literature [10]: harmonics are negligible for high static fields; their amplitude relatively to that of the fundamental increases with decreasing static fields. The non linear behaviour of the force is at its highest as shown at the right top of Figure 5, for a low static field and a high dynamic field.

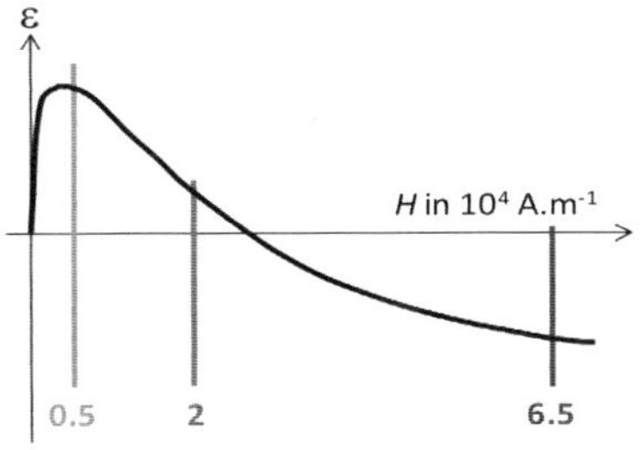

Figure 4: Steel magnetostriction curve, used for the examples of magnetostriction frequency spectra; the three static field amplitudes used are shown.

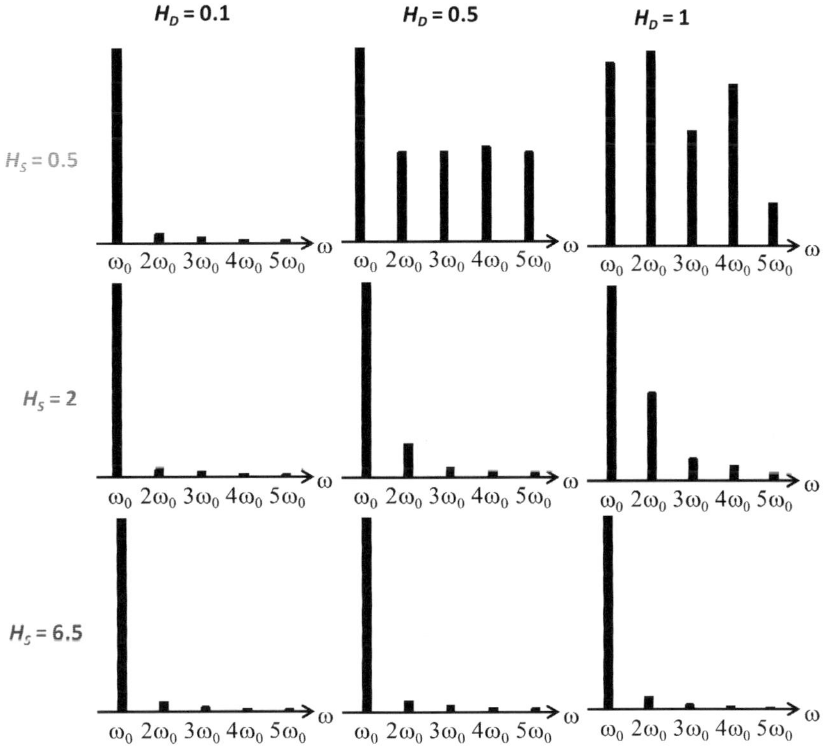

Figure 5: Normalized spectra of the magnetostriction force for different amplitudes of static and dynamic magnetic fields. Amplitudes in 10^4 A.m^{-1}.

The described algorithm allows one to compute frequency spectra associated to the magnetostriction force, extended to deal with arbitrary value of the bias field. This is a key point of our modelling approach since a high bias field assumption is generally made in existing theoretical modelling of magnetostriction. Moreover, the modelling relies on the knowledge of only one material input data, the macroscopic magnetostriction curve, which is easily measured experimentally.

3. Generation of guided waves by an EMAT in a ferromagnetic plate

3.1. Transformation of body forces into equivalent surface stresses

Thanks to the developments made in the previous section, the Lorentz and magnetostriction forces are calculated as frequency spectra in the volume of the part. These body forces are now considered as source terms in the equation of motion describing the elastic wave generation and propagation in the part. This is classically achieved by means of a convolution integral over the part volume of the source with the elastodynamic Green's tensor for the problem in hands.

By integrating the Lorentz force over the depth, that is, in the direction normal to the surface where the EMAT is applied, the volume integral can be transformed into a simpler surface integral [3,11]; this leads to the definition of a distribution of sources of surface stresses which approximately radiates the same wavefield as that radiated by the body force distribution. Proceeding to such a transform is computationally more efficient; moreover, the equivalent surface stress can be readily used in various existing models of wave radiation.

Unfortunately, such a simple integration scheme has been shown (Thompson [1,12]) to lead to a non-physical result when applied to the magnetostriction force, since the depth integral of this force vanishes. A more complicated integration scheme has been developed by Thompson [12] to overcome this difficulty. In two previous papers published in this series [6,7], Thompson's method has been reused. Final results in Thompson's paper were partly mistyped; we worked out the whole derivation of this transformation in details; this is rather technical and long and out of the scope of the present paper. The transformation may be found in [13] to which interested readers are referred.

3.2. Comparison with results from the literature

As far as Lorentz and magnetostriction forces are concerned, persistent discussions on their relative amplitude were found in the literature. In a recent joint paper by Ribichini, Nagy and Ogi [8], the magnetostriction force computed by Ribichini using a finite element method is compared to that computed by Hirao and Ogi [2] by means of an analytical expression of the force. In [2], boundary conditions of free surface were not taken into account. In the joint paper [8], the magnetostriction force for generating SH waves was considered and Ogi admitted that his results were wrong.
For validation purposes, the same case has been treated using our own modelling approach, consisting in computing the equivalent surface stress of the magnetostriction force; our results are compared to the various results given in [8]. As shown by Table 1, two or three orders of difference are found between Hirao and Ogi's model [2] and Ribichini's results. Our own results are of the same order as Ribichini's finite element results.

Table 1: Comparison of the order of magnitude of equivalent magnetostriction surface stress given by Hirao and Ogi [2], Ribichini *et al.* [8] and computed by the present method.

	Calculation method of the force	Magnitude of equivalent surface stress
Hirao and Ogi [2]	$\boldsymbol{\sigma}_{MS} = \mathbf{e}\,\mathbf{H}$	10^1 to 10^2
Ribichini *et al.* [8]	finite element	10^{-1}
present method	$\begin{cases} \boldsymbol{\sigma}_{MS} = \mathbf{eH} \\ \boldsymbol{\sigma}_{MS} \cdot \mathbf{n} = 0 \end{cases}$	10^{-1}

4. Results and discussion

Combining the calculation of both the Lorentz and the magnetostriction forces, their transformation into equivalent surface stresses and two models for predicting guided waves radiated by surface stresses (SH [6] and Lamb waves [7]), the modal amplitudes of guided waves generated by an EMAT into a ferromagnetic plate can be computed.

Here, an example of SH modal amplitude variation with the excitation frequency is given. The configuration used is shown on Figure 6 (left) and consists in a meander-coil associated to a tangential static magnetization. The 3-mm-thick plate is made of iron, characterized by the magnetostriction curve given by Hirao and Ogi [2] shown on Figure 6 (right).

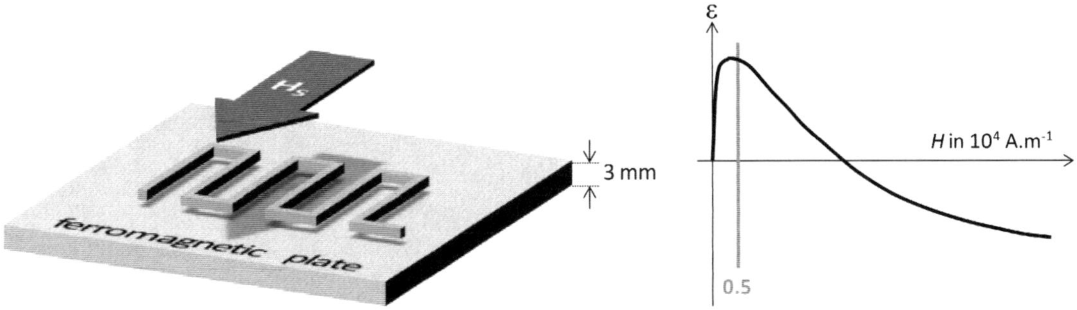

Figure 6: Left: EMAT configuration for generating SH waves. Right: magnetostriction curve used.

The modal amplitude of the first SH mode, SH0, is studied as function of the frequency for two current intensities (a low one of 40A and a high one of 400A). The left graph of figure 7 is computed for the low intensity of 40A, the right for the high intensity of 400A. Only the magnetostriction force is considered. Two different kinds of results are shown and superimposed on the same figure 7.

The first series (dark blue circles) represents the SH0 modal amplitude of the fundamental component of the magnestostriction force, over a frequency range from 0.1 to 1.0 MHz (note that the whole range is below the first cut-off frequency of SH guided waves in the plate considered).

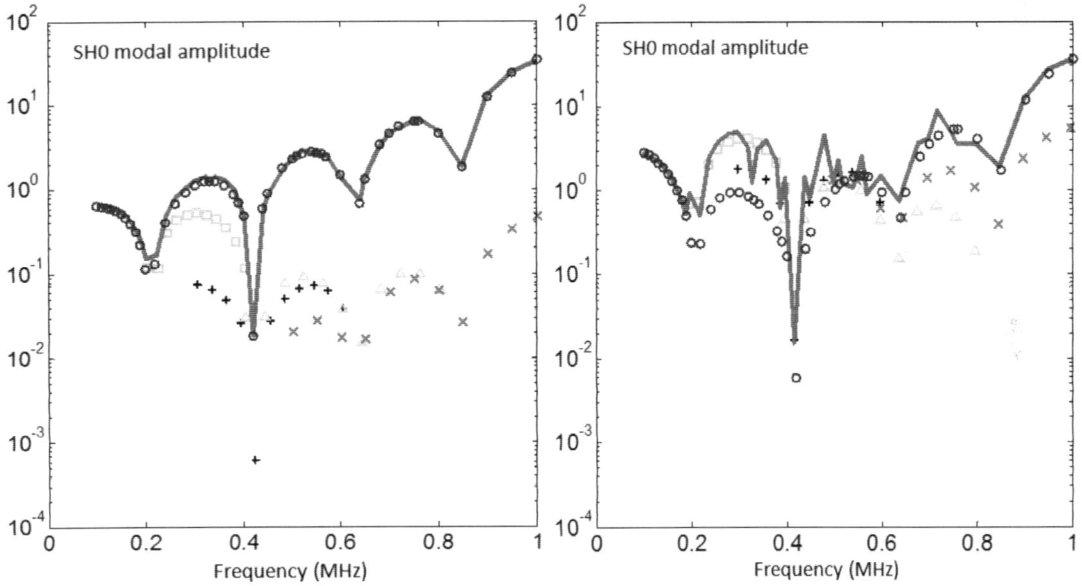

Figure 7: Left: 40A intensity; right: 400A. O: SH0 amplitude due to the fundamental component of the magnetostriction force in the range [0.1 -1] MHz. □ (resp. +, △, x,) contribution of 1st (resp. 2nd, 3rd and 4th) harmonic to the SH0 amplitude for a fundamental excitation in the range [0.1 - 0.2] MHz. Red curve: total amplitude of the SH0 mode for a fundamental excitation in the range [0.1 - 0.2] MHz.

The second series of results represents the SH0 modal amplitude of the first four harmonic contributions of the magnetostriction force, considering a fundamental excitation frequency varying in the 0.1-0.2 MHz frequency range. Therefore, the frequency range covered by these harmonics varies from 0.2 to 1.0 MHz. The various contributions are displayed separately (see figure caption). For the same excitation ([0.1-0.2] MHz), the fundamental and harmonic contributions are summed and the resulting amplitude is displayed as a red line.

In both graphs, a log-scale is used for the wave amplitude. Comparing the first series (blue circles) to the last (red line) allows us to quantify the effect of harmonic contributions. For a low intensity (left), both series almost superimpose: harmonics make a negligible contribution in this case. This is easily seen by looking closely at the individual contributions of the various harmonics; only the first harmonic in the range 0.2-0.4 MHz (green squares) slightly modifies the overall amplitude. A totally different behaviour is seen for high intensity (right graph). All over the frequency range of harmonics [0.2-1.0] MHz, contributions of the first four harmonics clearly cannot be neglected.

These results, obtained in a rather simple configuration, exemplify the ability of our approach to predict contributions of magnetostrictive forces in a wide variety of cases in terms of relative amplitude of static and dynamic magnetic fields. Our approach makes it possible to choose arbitrary values of bias field and of current intensity while most works make the restrictive assumptions of high bias field and low current intensity, the main reason for such restrictions being the difficulty to model the generation of harmonics by magnetostrictive effect.

Conclusions

A model has been proposed to compute the elastodynamic source terms associated to the magnetostriction force generated by an EMAT in a ferromagnetic part. Compared to existing semi-analytical models of magnetostriction, the present model can deal with arbitrary values of bias field and of current intensity thus with arbitrary EMAT design. This capability results from the specific development of a method to predict the complex-valued amplitude of contributions associated to harmonic frequencies of the fundamental excitation frequency. These contributions are negligible if high bias field and low current intensity are concerned, as shown by our results and as it is well established. Contributions from harmonics can however surpass that of the fundamental, as exemplified by our results dealing with a high current intensity. The present model can therefore help designing new EMAT for new applications.

Thanks to a mathematical method detailed in [13], the Lorentz and magnetostrictive body force distributions can be transformed into surface stress distributions. Finally, combining the calculation of force spectra with the transformation method of force into stress, the contributions of the fundamental excitation frequency and of its harmonics for both the Lorentz and the magnetostriction forces to any propagative mode (guided, bulk and surface waves) can be calculated.

Future work will include the calculation of non linear magnetic fields induced in the material. Accounting for non linear permeability shall increase the accuracy of the magnetic field calculation, thus, of the input data of the present model.

References

[1] Thompson R B 1978 *I.E.E.E. Trans. Son. Ultrason.* **SU-25** 7–15
[2] Hirao M and Ogi H 2003 *EMATS for Science and Industry: Noncontacting Ultrasonic Measurements* (Boston: Kluwer Academic Publishers)
[3] Thompson R B 1990 Physical principles of measurements with EMAT transducers *Physical Acoustics* vol 19 (New-York: Academic Press) chapter 4 pp 157–200
[4] Lhémery A 1994 *J. Acoust. Soc. Am.* **96** 3776–86
[5] Ditri J J and Rose J F 1992 *J. Appl. Phys.* **72** 2589–97
[6] Rouge C, Lhémery A and Ségur D 2012 *J. Phys.: Conf. Ser.* **353** 012014
[7] Rouge C, Lhémery A and Aristégui C 2013 *J. Phys.: Conf. Ser.* these proceedings.
[8] Ribichini R, Nagy P B and Ogi H 2012 *NDT&E Int.* **51** 8–15
[9] Bozorth R M 1964 *Ferromagnetism* (Hoboken: Wiley-Interscience)
[10] Laguerre L, Aime J-C and Brissaud M 2002 *Ultrasonics* **39** 503–14
[11] Kawashima K 1984 *I.E.E.E. Trans. Son. Ultrason.* **SU-31** 83–94
[12] Thompson R B 1980 *J. Nond. Eval.* **1** 79–85
[13] Rouge C, Lhémery A and Ségur D 2013 *J. Acoust. Soc. Am.* **134** 2639-46

AUTHOR INDEX

Institute of Physics
Dirac House, Temple Back
Bristol BS1 6BE UK

proceedings.com